NANOVISION

NANOVISION

ENGINEERING THE FUTURE

Colin Milburn

Duke University Press

Durham & London

2
0
0
8
.
.
.

© 2008 DUKE UNIVERSITY PRESS

All rights reserved
Printed in the United
States of America on
acid-free paper

Designed by Amy Ruth Buchanan
Typeset in Scala by Tseng Information
Systems, Inc. Library of Congress
Cataloging-in-Publication data appear
on the last printed page of this book.

CONTENTS

ACKNOWLEDGMENTS

> We have technology to manipulate matter right down to
> the molecular level. This is an extraordinary ability, think
> of it! And yet some of us here can accept transforming
> the entire physical reality of this planet, without doing a
> single thing to change ourselves, or the way we live. . . .
> And so I say that among the many things we transform
> on [this planet], ourselves and our social reality should
> be among them. We must terraform . . . ourselves.
> —Kim Stanley Robinson, *Red Mars*

Like the future itself, this book has been a while in the making. My fascination with the dawning era of nanotechnology began during the mid-1990s, and in the intervening years my thinking on the topic of the very, very small has benefited enormously from conversations with many friends and colleagues about science, fiction, and the absolute weirdness of the shape of things to come. The terraforming of the self begins with such conversations, in the interface.

Thanks to Brian Attebery, Richard Bell, Bruce Clarke, Seeta Chaganti, Joshua Clover, Lucy Corin, Frances Dolan, Eve Downing, Joseph Dumit, Margaret Ferguson, Marjorie Garber, Michael Fisher, Valerie Hanson, Jochen Hennig, Arne Hessenbruch, Sarah Jansen, Matthew Jones, Despina Kakoudaki, Mario Kaiser, Monika Kurath, Brooks Landon, Andreas Lösch, Elizabeth Lyman, Kamila Lis, Lis Maguda, Timothy Morton, David Mulrooney, Brigitte Nerlich, Alfred Nordmann, Charles Ostman, Laura Otis, David Robertson, Kim Stanley Robinson, Joachim Schummer, David Simpson, Sha Xin Wei, Mark Solovey, Christopher Toumey, Ana Viseu, Ann Weinstone, and Michael Ziser for ideas and inspirations along the way.

Most of the chapters in this book appeared originally as conference presentations, and I appreciate the generous feedback I received from audiences at the 2004 meeting of the History of Science Society; the UCLA Electronic Literature Organization's 2004 conference "Self-Organizing Systems: rEvolutionary Art, Science, and Literature"; and the 1999 colloquium "Science and High Technology in the Silicon Valley" at Stanford University. I also profited tremendously from discussions with fellow participants in the "Imaging Nanospace — Bildwelten der Nanotechnik" workshop at the University of Bielefeld's Zentrum für interdisziplinäre Forschung (ziF) in 2005.

For their help with some of the images reproduced in this book, I thank Charlotte Bolliger, Sara Delekta Galligan, Kelli Gaskill, Steve Gerard, Wolfgang Heckl, Eileen LaManca, Colleen Leyden, John Mamin, Mike Ross, and Richard Vine.

I am particularly indebted to Neil Badmington, Robert Brain, Alisa Braithwaite, Luis Campos, Tyler Curtain, Olivier Darrigol, Mark Jerng, Nadine Knight, Monica Lewis, David Munns, Sharrona Pearl, Gillian Paku, Laura Thiemann Scales, Amanda Teo, Marga Vicedo, and Carol Wald for their enduring support and for reading numerous fragments of this project over the years. Two anonymous referees for Duke University Press carefully read the entire manuscript, and I deeply appreciate their insights and suggestions. Thanks also to Ken Wissoker, Courtney Berger, Bill Henry, Pam Morrison, and all the others at Duke University Press who have made this book possible.

John Bender, Mario Biagioli, Peter Galison, Stephen Greenblatt, N. Katherine Hayles, Barbara Johnson, Timothy Lenoir, Seth Lerer, Everett Mendelsohn, Katharine Park, Patricia Parker, and Ann Wierda Rowland have helped me in more ways than I can even begin to count. My teachers and my friends — thank you, all.

Some of the research for the project was supported by a grant from the Harvard University Humanities Center, for which I am sincerely grateful. Earlier versions of chapter 1 and chapter 4 appeared in *Configurations* 10 (2002): 262–95, and *New Literary History* 36 (2005): 285–311, respectively. I thank the Johns Hopkins University Press for granting permission to reproduce this material here.

And finally, my warmest thanks to the loved ones who have been with me through the passing of time, sharing in the greatest adventures as we move all together into unknown futures. My whole family has been a source of endless encouragement and strength; without them, this book would not exist. I would especially like to thank Peter Holman, Elizabeth Lee, Douglas

Milburn, Dustin Milburn, Renee Milburn, Jonah Mitropoulos, and James Robertson for being there every step of the way. And not to be forgotten, those nonhuman companions whose affection meant the world to me during the many years it took to complete this book, who are still loved even though some are now gone: Noli, Zipper, Pockets, Besame, Rosemary, Carmen, and Figaro.

INTRODUCTION

THE SINGULARITY OF NANOVISION

> Additional view-windows kept popping up as the nano-
> machines multiplied. . . . Only a minute had elapsed, but
> the world felt different. Human history had changed for
> good.
> —Rudy Rucker, *Postsingular*
>
> The universe grows smaller every day.
> —*The Day the Earth Stood Still*

It's coming. Or rather . . . it's here.

In 1993, the mathematician, computer scientist, and science fiction writer Vernor Vinge prophesied the end of human history. Once again. It's the end of the world as we know it . . . and not for the first time. So this would appear to be nothing new. And yet . . .

Looking at the rapid acceleration of technological progress over the course of the twentieth century, Vinge observed an exponential growth curve in the development of computational systems, bioengineering capabilities, human-hardware interfaces, and machinic intelligence, all of which seemed to suggest that a cataclysmic takeoff of technological complexity would likely occur sometime between 2005 and 2030. The growth curve of technoscientific progress would rise asymptotically toward infinity, and nothing would thereafter remain the same. Vinge termed this point in the future "the Singularity." The impact of these technoscientific changes on human society would be so overwhelming as to constitute not simply a new era in history but the onset of a new reality entirely, a new mode of being. The Singularity, Vinge writes, will be "a point where our old models must be discarded and a new reality rules, a point that will loom vaster and vaster over human affairs until the notion becomes a commonplace. Yet when it finally happens, it may still be a great surprise and a greater unknown." Indeed, before we are perhaps even aware of it, "we will be in the Posthuman era."[1]

This idea of "singularity" comes from mathematics and astrophysics, where it indicates the point at which a function rockets to infinite value, or where the fabric of space-time collapses to a point of infinite curvature, such as general relativity predicts will occur inside a black hole. Within a space-time singularity, the established rules of physics no longer apply; as the physicist Stephen Hawking puts it, "At the singularity, general relativity and all other physical laws would break down: one couldn't predict what will come out of the singularity."[2] In adapting this scientific concept for futurological purposes, Vinge came to understand the technological Singularity as an "edge of change" in human evolution, marking our entry into "a regime as radically different from our human past as we humans are from the lower animals. This change will be a throwing-away of all the human rules, perhaps in the blink of an eye—an exponential runaway beyond any hope of control" ("TS," 89). The human species would transform so utterly during the Singularity as to be alien from its current condition.[3] But understanding the nature of this change, or the features of this new "regime," would be prevented by the very acceleration of the change itself. Because the Singularity "involves an intellectual runaway, it will occur faster than any technical revolution seen so far" ("TS," 90). In the same way that a singularity in a mathematical function blocks extrapolation of the curve beyond the point where it shoots upward toward infinity, or that a black-hole singularity in space traps light and prevents us from seeing beyond its event horizon, the technological Singularity blocks our ability to see the future. It cannot be extrapolated by past experience or by scrutinizing current tends; it remains "unseen" precisely because it is so different from any other era of technological change that has been "seen so far."

The very question of seeing is as much at stake in the technological Singularity as it is in the astrophysical singularity of the black hole, for they are both points of blindness where the human conceptual apparatus—dependent, at least metaphorically, on light, sight, and vision—cannot penetrate. According to Vinge, the Singularity is a moment of darkness, a point that occurs "in the blink of an eye," a spot in time where we literally cannot see. Those who try to rigorously understand the consequences of technological change for human culture—such as science fiction writers and futurologists—increasingly find their visionary abilities curtailed by the "unknown" of the Singularity, their speculations sent careening asymptotically in multiple directions at the event horizon of this black hole in history. The Singularity blocks prediction and visionary speculation; it is "an opaque wall across the future" ("TS," 90). Consequently, from this side of the Singularity, even the "most diligent extrapolations [have] resulted in the unknowable" (90). The Singularity becomes a pure event, cleanly separating the past from the future, a cleavage "that we cannot prevent," for "its coming is an inevitable consequence of humans' natural competitiveness and the possibilities inherent in technology" (92). And because we cannot see beyond this inevitable transformative event, all we can know is "how essentially strange and different the Posthuman era will be" (95).

Vinge's own fictional efforts to characterize the technological events surrounding the Singularity repeatedly evaporate into mystery.[4] The characters in his novel *Marooned in Realtime* (1986), for example, travel in temporal suspension to a historical point after the Singularity has already occurred—essentially, they unwittingly sleep through the Singularity. The sleepers wake to discover Earth's human population and all its technological traces vanished. No amount of study will reveal the truth of the disappearance, the nature of Singularity. It is a pure point of undecidability. Assembling pieces of historical evidence from before and after this void in time, the characters' best guess is that "humankind and its machines became something better, something . . . unknowable." But "if technology had transcended the intelligible [and] . . . if minds had found immortality by growing forever past the human horizon," then this "human horizon" would itself mark the limit of specularity and speculation. The limit of the intelligible functions, therefore, as a mirror: "The Singularity was a mirrored thing."[5] Looking at it, speculating on it, only reflects the human past back to itself: alternative futures remain invisible, veiled by the mirror. We are blind to the beyond.

Scientific and philosophical dialogues about the Singularity have repeatedly emphasized this characteristic blindness. The critical theorist Damien Broderick describes the onrushing rapid acceleration of machine intelligence as "the edge of a technological Singularity, the place when the future starts to go completely opaque." Once we pass the edge, the "future is going to be a fast, wild ride into strangeness," slipping by in "(historically speaking) the blink of an eye."[6] This opaque, estranging future that meets us suddenly when our eyes are closed appears as a violent scission, a slicing of time by the cutting edge of complexity. With an ironic wink to millenarian clichés, the computer scientist Ray Kurzweil has announced that "the Singularity is near,"[7] and he insists that we are coming upon this cutting edge sooner than we think: "We are entering a new era. I call it 'the Singularity.' It's a merger between human intelligence and machine intelligence that is going to create something bigger than itself. It's the cutting edge of evolution on our planet."[8] This cutting edge makes a break in history, a division that separates our knowable past from the impossibly strange future, as if rupturing or puncturing our very eyes as they peer into the distance.

Hans Moravec—a roboticist well known for his prognostications on the evolution of machine intelligence and the theory of "uploading," or the transference of human mind into computer code[9]—has written: "If there is a singularity, it's kind of natural to divide time into BS (the negative times before the singularity) and AS (the strange times afterwards)."[10] Our living history recedes into absolute negativity (or, indeed, into "BS") relative to the force of the unknown future, whose strangeness is mathematically infinite and is perceived thus as an absolute positivity. It is as if this singular blade has already fallen onto the Cartesian grid of human temporal existence, just ahead of us, but also just out of sight, for we would seem to have already dropped into the abyss of obsolescence, erased by a future fundamentally outside our peripheral vision. Indeed, as the physicist and science fiction writer Gregory Benford suggests, most of us will never see the Singularity; we won't be aware of it even when it arrives, for while some sectors of humanity will whisk across this transition, most will never *see* it happening because "those in the Singularity will be beyond view, anyway."[11]

Max More, a transhuman theorist and founder of the futurological Extropy Institute, suggests that various scenarios for massive technical change are possible. Some indicate radical severance and discontinuity, others promise a rapid but continuous burst into strangeness, but all find the future imperceptible from within our stygian hole of mere humanity:

> This Singularity includes the notion of a "wall" or "prediction horizon"—a time horizon beyond which we can no longer say anything useful about the future. The pace of change is so rapid and deep that our human minds cannot sensibly conceive of life post-Singularity. Many regard this as a specific point in the future, sometimes estimated at around 2035 when AI and nanotechnology are projected to be in full force. . . . The more that progress accelerates, the shorter the distance measured in years that we may see ahead. . . . Singularity [can also be] seen as a *surge* into a transhuman and posthuman era. . . . In Singularity as Surge the rate of change need not remotely approach infinity. . . . It would be a historically brief

phase transition from the human condition to a posthuman condition of agelessness, super-intelligence, and physical, intellectual, and emotional self-sculpting. This dramatic phase transition, while not mathematically instantaneous, will mean an unprecedented break from the past. Second, since the posthuman condition (itself continually evolving) will be so radically different from human life, it will likely be largely if not completely incomprehensible to humans as we are today.[12]

The envisioned Singularity, whether characterized as a "wall" or a "surge," remains a decisive event that divides history itself so cleanly, so cataclysmically, that what we know as "the human" cannot see past its own closure. It is a failure of humanism itself. As Vinge writes, "The problem is not simply that Singularity represents the passing of humankind from center stage, but that it contradicts our most deeply held notions of being" ("TS," 94).

We cannot see past the Singularity because to do so would involve an entirely different way of seeing, a new epistemological orientation toward the world, a new thinking of being that is no longer the perspective of the human, but instead that of the posthuman, the postbiological, the machinic, the cyborg, the networked, the uploaded, the synthetic, the schizophrenic, the alien, the monstrous, the wired, and the weird. Outside the spaces delineated by humanist sensory capabilities, this technologically involved perspective of radical alterity, as N. Katherine Hayles has written, would see "no essential differences or absolute demarcations between bodily existence and computer simulation, cybernetic mechanism and biological organism, robot teleology and human goals."[13] Precisely because the Singularity, or "the Spike," as Broderick prefers to call it, "could change everything utterly, in ways too ruinous and horrifying to regard with merely human gaze," we require other forms of perception unhampered by epistemic limitations of the visible or intelligible.[14] Indeed, the only way to see through the looking glass, the only way to glimpse the posthuman future across the opposite side of the wall, would be to render the stasis of human vision into perceptual motion, through the active involvement of the observant body in technological events, in a real physical passage through and across the singular limits of knowledge. Travel or tunnel through, or carry across in a surge, as Vinge tells us, and "then it's *you* . . . who will understand the Singularity in the only possible way—by living through it."[15] Amazingly enough, it turns out that the technological events through which we might involve ourselves bodily and intelligibly in this "living through" of the invisible future are already at hand.

Among all the technoscientific developments invoked by theorists of Singularity as components in the technological eventstream leading inexorably toward the altered reality of the posthuman era—including artificial superintelligence, genetic engineering, artificial life, evolutionary robotics, cloning, synthetic biology, and ubiquitous computing—perhaps none has seemed as prominent or as promising as nanotechnology. Nanotechnology is the engineering of material structures and functional systems at the scale of nanometers (billionths of a meter), where individual molecules and atoms become objects of manipulation. Nanotechnology strives to take advantage of unique properties of matter at this molecular scale, for example, the phenomenon of "quantum entanglement" that might enable quantum computing technologies, or likewise the hypothetical phase state of "machine-phase matter," whose volume would be filled entirely with active molecular machinery.[16] Nanotechnology also hopes to develop assembly processes for manufacturing microscale and even macroscale structures "from the bottom up," making conductive materials, motors, biomimetic organelles, and computational processors—even sophisticated robotic systems (figure 1)—by maneuvering individual atoms. In other words, nanotechnology dreams of engineering every aspect of our material reality, precisely fashioned and designed at the limits of fabrication, one atom at a time.

The possibilities opened by the capability to restructure and rearrange matter at the nanoscale are immense, making speculation on the future an almost inherent aspect of thinking about nanotechnology in the first place. For if nanotechnology enables us to program matter as we would program software, then the world itself can be transformed, our lived realities made completely malleable, guaranteeing that the future will be radically and immeasurably different from the present. While the precise nature of this incomprehensible difference remains to be seen, or lived through, according to Singularity theorists, it is already evident that simply by peering into the resources and capabilities of nanotechnology, we suddenly find ourselves "accelerating into a future that's literally beyond today's imagination because its complex weaving of the known and the as-yet-unknowable evades the best calculations we can make." So ready or not, "Nano will take us, will *fling* us, into the Spike."[17]

Nanotechnology as an emergent technoscientific field is actively and rapidly developing across multiple scientific disciplines, from chemistry, physics, and biology to computer science, materials engineering, and systems theory. It is a fundamentally multidisciplinary endeavor that draws its research techniques, theoretical approaches, and laboratory apparatus from

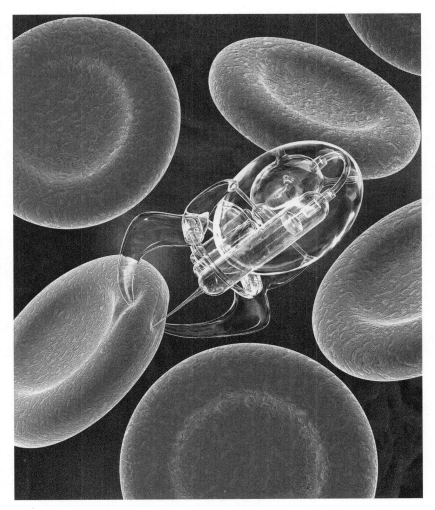

1. "Nanoprobe" (2002), by Coneyl Jay. A diamondoid syringe-bot travels through the body, delivering drugs to red blood cells. Originally entitled "Nanotechnology," this conceptual image received the Visions of Science Award in 2002 (sponsored by the *Daily Telegraph* and the pharmaceutical company Novartis). © Coneyl Jay/Photo Researchers, Inc. Reproduced with permission.

many different traditions and regions of scientific specialization.[18] Its possibilities are being explored both experimentally and speculatively across academia, industry, and popular culture. The term "nano-technology" first appeared in a 1974 article by Norio Taniguchi, a professor at the Tokyo Science University, although the conceptual origins of the field have often been traced to the hopeful prognostications of the Nobel Prize–winning physicist Richard Feynman in the late 1950s.[19] Yet nanotechnology only fully emerged as a research program with monumental implications for the human future in the writings of K. Eric Drexler during the 1980s and 1990s. Along with his several important technical articles and monographs, Drexler's *Engines of Creation: The Coming Era of Nanotechnology* (1986) inspired legions of scientists and techno-enthusiasts with the feasibility and the consequences of developing this new science. Drexler's writings offered a grand picture of the future dramatically transformed by the advent of nanotechnology, even predicting something very like a Singularity in the wake of mature molecular manufacturing. Drexler's books drew widespread public attention to the possibility of nanoscale engineering, which for many years, and even until recently, was considered mere "science fiction" by many in the scientific community.

Since the 1980s, nanoscience has exploded across the world, attracting researchers from surprisingly different disciplinary backgrounds and from multiple technoscientific sectors. Ambitious legislation and funds for large-scale nanotechnology initiatives have recently been put in place by the governments of the United States, the European Union, Japan, the United Kingdom, China, Singapore, and many other countries. Large technology corporations and smaller start-up companies have announced nanotech R&D efforts, anticipating consumer products with "nano inside" sometime in the very near future. Already the prefix "nano" pops up all over popular culture. Television programs, novels, films, advertisements, comic books, and video games depicting nanotechnology and its implications appear nearly every day. The international hipness of nano even spills over to businesses that have little to do with the nanoscale as such (figures 2 and 3).

Nanotechnology, the science of the very small, has clearly become "big science." And though many of the research programs in academia, industry, medicine, and the military that make up the technoscapes of nanotechnology actively disavow any connection to the futurist imaginings of Drexler and his kindred "exploratory engineers"[20]—indeed, some nanoscientists have stridently suggested that Drexler's ideas about nanotechnology are not only impossible but dangerously misleading—it is nevertheless the case that virtually all sectors of nanotechnology research strongly maintain that the technical ability to manipulate, program, and engineer matter at the molecular level heralds staggering and unprecedented transformations for our world.[21]

The scientific agencies of the U.S. government foresee the development of nanotechnology leading to "the next industrial revolution," a massive "technological convergence" at the nanoscale that will restructure both the international economy and the human body itself.[22] The nanoscientist and Nobel laureate Richard Smalley has said, "There is a growing sense in the scientific and technical community that we are about to enter a golden new era. We are about to be able to build things that work on the smallest possible length scales, atom by atom, with the ultimate level of finesse. These little nanothings, and the technology that assembles and manipulates them— nanotechnology—will revolutionize our industries and our lives."[23] Everywhere we are told that nanotechnology is "the next big idea," and we are advised of "the big changes coming from the inconceivably small."[24] The world appears to tremble under the pressure of all this expectation, and our global societies are perched on the brink of immense technological revolution by virtue of all this hyperbolic rhetoric, this inflated "nano-hype."[25] Something REALLY BIG is on the horizon, largely unseen and essentially inconceivable, but do not doubt that it is coming.

Within this action-packed, adrenaline-pumping discourse of profound, cataclysmic, unprecedented transformations that could arise as direct consequences of nanotechnology, the Singularity seems to hover in the background as their culmination. Indeed, the very conditionality of these possible changes, this question of the "could," actually locates a conceptual singularity inside nanodiscourse itself. The very possibility that nanotechnology could change the world is a subjunctivity presenting itself as the event horizon of the unseen future, as the proximal limit of a future that cannot be known other than in its radical difference from what is present. This subjunctivity is a blinding, an incisive wound, made by the cutting edge of nanotechnological research as an internal and inherent feature of thinking the possibilities of nanotechnology.

As Mark Gubrud, a researcher in quantum computing, has said, "The concept of a singularity follows directly from the original concepts of molecular nanotechnology."[26] The imagination of nanotechnology would suggest a nearly infinite number of alternate futures made available by the ability to rebuild reality from the bottom up, one atom at a time—a superposition of futures emergent from all conceivable reconstructions of any

1,000 songs: Impossibly small. **iPod** nano

2. Apple iPod nano: "Impossibly Small." Although Apple's petite media player is far from nanoscale, its phenomenal market success has further escalated cultural enthusiasm for the prefix "nano." As this magazine ad from November 2005 implies, nano is now practically at our fingertips, within our grasp, and we touch the "impossible."

given material assemblage—defying our abilities of prediction and making the future increasingly uncertain with every advance toward actually achieving a mature nanotechnology. Moreover, given the possibility of molecular manufacturing, the geometric or exponential acceleration of technological complexity seems *already inevitable*. Several nanotech theorists have argued that as soon as nanotechnology begins to seem possible, its continued development becomes unavoidable because a technological imperative takes over beyond human control.[27] This technological imperative would drive us insistently to a moment in the future beyond which we cannot see—the blind spot of Singularity—owing to our physical and conceptual limitations relative to our own rapidly developing technology. As Gubrud puts it, once you have nanotechnology, "that could lead to a singularity, because the rate of technological progress would be set by technology, rather than the speed at which people work. . . . We are facing in the next few decades a time of very great technological change, primarily driven by nanotechnology and microelectronics."[28] According to the nanotheorist and science fiction writer John Robert Marlow, nanotechnology bears forth "the sound of inevitability" because so many industrial, corporate, military, governmental, economic, cultural, and scientific incentives already exist for its continued progress: "Given all of this—can nanotechnology *not* happen?"[29]

Nanotechnology thus becomes the most recent in a long assembly line of mechanical developments—from the factory system to cybernetics to AI— envisaged to become autonomous and self-evolving, driving the modern era through an uncontrollable technological determinism.[30] This sense that technology sets the pace of its own development therefore undergirds the rolling road to Singularity, for Vinge writes that advances in technological automation are so attractive on every level of social organization—domestic, industrial, artistic, economic, military, and so forth—that our progress toward a moment when "greater-than-human intelligence drives progress" is already destined, we have already lost ourselves to the acceleration, and

3. "Nano Energy Underclothes." In the streets of Hong Kong, a 2006 billboard displays the mysterious pleasures of "nano energy," the intimate and invigorating touch of the infinitesimal against the skin. Photograph by Richard Vine. Reproduced with permission.

納米能量內衣褲

Nano energy underclothes

there is no slowing down: "If the technological Singularity can happen, it will" ("TS," 89, 91).

It would seem, then, that we are inevitably made aware of Vinge's "opaque wall over the future" as an immediate consequence of thinking nanotechnology. In other words, the imagination of nanotechnology itself constructs this wall as its own internal limit, discovering that its potential is so imperative that a nanotechnology future is rendered already inevitable, but also that its potential is so vast that a nanotechnology future is rendered equally uncertain and indeterminate. The imagination of nanotechnology creates its own blindness, credited with an ability to change the future so utterly as to make that future unimaginable within the limits of human perception. We are blinded by our own efforts to conceive "the big changes coming from the inconceivably small." The Singularity appears as the edge of nanotechnological speculation, the barrier across which the present cannot cross, or even see.

And yet, paradoxically, some theorists simultaneously credit the imagination of nanotechnology with a visionary perspicacity that surpasses the blind spot of the Singularity. Kurzweil has written that "we cannot easily see inside the event horizon with certainty. . . . Nevertheless, just as we can draw conclusions about the nature of black holes through our conceptual thinking, despite never having actually been inside one, our thinking today is powerful enough to have meaningful insights into the implications of the Singularity."[31] Indeed, such impossible insights seem to be made possible by the conception of nanotechnology and its kindred fields at the cutting edge of science. Our thinking today is rendered powerfully and prophetically insightful by virtue of the rapid convergence of numerous technosciences at the nanoscale, the momentous conflation of biotech, cognotech, infotech, and more under the blooming sign of "nanotechnology."[32] The theorists, futurologists, scientists, and science fiction writers who situate their gaze within the parameters of a nanotechnological way of seeing may encounter a singular wall over the future—a black hole, a constrictive passageway, or a surge into the future that lets no light escape—but in encountering it, these visionaries see through it. Which is to say that in seeing the molecular world—or rather, the nanoworld—we now also see the nanofuture: "What [nanotech] did . . . was shatter the event horizon."[33]

At the very moment of describing the opacity of the Singularity, Vinge writes that the advent of nanotechnology simultaneously "provided spectacular insights about how far technical improvement may go" ("TS," 91). The thinking of nanotechnology provides surprising visions—"spectacular insights"—penetrating glances into the smallest limits of molecular space that therefore open out the reaches of the future. Peering inward, they extend to the lengths of the possible ("how far technical improvement may go"). These visions into nanotechnology itself, these "insights" that see out through the far technological future—these openings across the Singularity as bursts of internal visibility that make "spectacles" of themselves—enable us to see in their being seen; they provide us with visionary enhancement precisely because they extend our limited sensorium across a techno-theoretical prosthesis, a conceptual pair of nanotech goggles or spectacles. Spectacular insights provided by nanotechnological thinking thus extend our ability to think nanotechnologically, inward and outward, across the limit of technological development. As Yoshio Nishi, a director of the Stanford University Nanofabrication Facility, puts it: "Nanotechnology is the tunnel we can take to get past that barrier. . . . There will be many engineering challenges but the path is there and we just need to keep following it. This is not science fiction."[34] The insights of nanotechnology in some way thus evade the opacity of the Singularity. Looking into itself, nanotechnology looks outward from blindness—and sees otherwise.

Nanotechnology entails a way of seeing, a perspectival orientation to the world, that operates through a productive dynamic of blindness and insight.[35] It produces a blind spot, a wall, a veil, a black hole, or a barrier and therein discovers a scission—between present and future, between human and posthuman, between science and science fiction. But at the same time, even in discovering its own blindness, it sees through it toward the beyond. It breaches the wall, breaks the barrier, lifts the veil, and voyages into the black hole. It is a way of seeing that lyses the membrane between the technological present and the nanotechnological future.[36]

I call it *nanovision*.

By tracing the cultural history of nanotechnology and examining its rhetorical, textual, and imaging practices, by looking at the structure of nanotechnological experimentation in both science and fiction, this book puts forward a theory of nanovision as a seriated movement of specularity and speculation that organizes the technoscapes and dreamscapes of nanotechnology. Nanovision is a perceptual apparatus endemic to the era of nanotechnology, atomizing our world only to perform its molecular reconstruction, envisioning ultimate limits only to speculate on their outside, fabricating barriers only to tunnel through them, projecting opaque walls only to find in the very project an excuse or an opening for spectacular insight. Through engineering a series of epistemic and rhetorical dichotomies within its discursive domain and simultaneously rupturing their conceptual separa-tions—the dialectics of nanovision—nanotechnology makes a radically dif-ferent future possible even now. For within its assemblages of texts, images, narratives, technical artifacts, and scientific instruments, nanotechnology gives rise to this way of seeing that makes the otherwise unthinkable ex-terior of Singularity—the end of technological advancement from the per-spective of human history—available to our imagination.

Which is not to say that nanovision simply escapes or disappears into the posthuman future; on the contrary, as we will see, nanovision depends on animating a productive dialogue and conflict between presentism and futur-ism, between humanistic thought and its other. But in negotiating between the conflictual elements of its own discourse, nanovision sees its blindness and therein discovers traces of alterity. In noticing its own internal singu-larities—or in discovering the Singularity proper—it opens to unknowable futures, brings those futures of endless possibility into the present, and thereby builds the epistemological conditions for inhabiting the future as such. It does not escape, but it opens to its beyond.

This produces a ceaseless back-and-forth motion, a sort of Fort/Da game of speculation and recall simulating the surface tension between inside and outside; and even when extending lines of flight from within the enclosed worldview of contemporary technoculture, nanovision retains certain limi-tations of its present condition. Its blindnesses are those of humanism and human perspectivalism more generally. Jettisoning itself from linear his-tory and seeing the present retrospectively and already nostalgically from the perspective of the future, nanovision would seem to be a profoundly postmodern development. But at the same time, its anterior knowledge of the future depends on its technological determinism, its insistent echo of the "sound of inevitability." It evidently enacts a grand teleological narrative of future history that appears retrenchant in the face of postmodernity's notorious "incredulity toward metanarratives."[37] For in discovering the Sin-gularity, nanovision appears to replicate a humanist and even religious tele-ology of the "end of man," the eschatology of the world and the apocalyp-tic transcendence of being.[38] Appropriately, then, the Singularity has been termed the "theology of the ejector seat" and "the rapture of the geeks."[39]

But even in animating this confrontation between humanist and post-humanist metaphysics at the site of the Singularity, nanovision discovers its own blindness and works through it, deconstructing and reconstructing the historical and metaphysical framework on which it depends. Nanovision encounters the paradox of announcing simultaneously the unknowability of the future and the inevitability of the future, and within this paradox it unfolds an endless process of transverse movement that does not escape but manages, in motion and action, in the involvement of human perception with technological otherness, to replace the static being of transcendent "rapture" with the participatory evolution of "becoming."[40] Max More has written of precisely this issue:

> As the near-universal prevalence of religious beliefs testifies, humans tend to attach themselves, without rational thought, to belief systems that promise some form of salvation, heaven, paradise, or nirvana. In the Western world, especially in millenarian Christianity, millions are attracted to the notion of sudden salvation and of a "rapture" in which the saved are taken away to a better place. . . . I am concerned that the Singularity concept is equally prone to being hijacked by this memeset. This danger especially arises if the Singularity is thought of as occurring at a specific point in time, and even more if it is seen as an *inevitable* result of the work of others. I fear that many otherwise rational people will be tempted to see the Singularity as a form of salvation, making personal responsibility for the future unnecessary. . . . Clearly this abdication of personal responsibility is not inherent in the Singularity concept. . . . I think those of us who speak of the Singularity should be wary of this risk if we value critical thought and personal responsibility.[41]

Observing the Singularity through something like a critical nanovision, More finds the posthuman "memeset" ripe for being "hijacked" by the memeset of rapturous religiosity—if, indeed, it has not always already been deeply inhabited by this theological structure. Discovering, then, the very limitations of seeing the Singularity through humanist eyes, More advocates "critical thinking," an insistent self-reflection and analysis, and a location of visionary perspective into the self, a refusal of seeing the technological Singularity or the nanofuture as the "*inevitable* result of the work of others," but rather as the concentrated involvement of ourselves in the technocultural process of becoming-posthuman. He does not blithely jump into the post-Singularity future, despite his evident desire to do so, but instead recognizes this very temptation and veers off, using this insight to propose a participatory making of the future—indeed, a "responsible" engineering of the future. Unlike a rapturous humanism where the body can so easily be discarded, nanovision would be located within the self, within the body, within "personal responsibility" as a perceptual and responsive engagement in becoming.[42] As Wil McCarthy puts it in his science fiction novel about nanotechnological singularity, *Bloom* (1998): "We can't ask things to happen by themselves; vision is transmuted to physicality through our hands, only."[43]

Nanovision thus performs a "techno-deconstruction" of the very structures of thought and embodiment through which it has come into being. By this I mean it challenges, questions, and revises the limits of human being at the level of metaphysics and imagination, as well as the level of corporeal materiality. For example, in Ben Templesmith's graphic novel *Singularity 7* (2005), the world is taken over and molecularly reengineered by a plague of alien nanotechnology: "They called it 'The Great Unravelling.' 4 billion people disassembled on a molecular level by the very air, swarming with nanites they simply breathed in. . . . Some that were left . . . they tried to fight back. But it was useless. How do you fight something that is in the very air? That deconstructs you on a molecular level?"[44] More than metaphor, this description perfectly condenses nanotechnology as a conceptual apparatus and a technical system of artifacts and instruments. Nanotechnology is "something that is in the very air." It swarms in the air even now; it infiltrates the zeitgeist. We begin to think it, with it and through it, even as its technical operations begin to take place in the world. And in thinking through it, indeed, nanotechnology "disassembles"; it "deconstructs you on a molecular level" (figure 4). Nanovision—this term for thinking through nanotechnology in its theoretical operations and material instantiations—carries out a techno-deconstruction, an unraveling, a desedimentation of human being. For nanovision animates the molecular tensions within humanism and the human body itself and works through these molar structures toward molecular modalities of becoming.

Nanovision's techno-deconstructive effects problematize the difference between the human and the extrahuman, opening the human to those nanotechnologies that "are in the air," be they machinic "nanites" per se or operational forms of technical nano-knowledge. Nanovision finds the point, the fissure, where presentist humanism fails, and it is at this critical failure that the possibility of posthumanism emerges, as a processual movement of self-othering.[45] This blind spot or limit within the scope of humanism is the very condition for becoming other than human: it is the fault line marking the trace of the inhuman within the human, of the future within the present, of the impossible within the possible—a critical failure of what is properly thinkable within human thought. Seeing what cannot be seen, discovering monumental historical changes that are both inevitable and un-

4. Ben Templesmith, *Singularity 7* (2005). In the scene of nanotechnological disintegration, human beings are graphically broken down and analyzed, "deconstruct[ed] . . . on a molecular level." © 2005 Ben Templesmith and Idea + Design Works, LLC.

known, nanovision makes the outside appear within the inside as a trace of absolute alterity. It sees, exposes, and produces the invisible future on this side of the Singularity—the future is presented, emerging from nanotechnological developments that have not yet happened (and perhaps never will) but, in being anticipated, enact change in the world. In bringing to light the traces of this unimaginable future inside the human present, nanovision thus finds the tunnel, the exit, the way out, releasing a flooding technological surge through the constricting sphincter of the Singularity—which is, therefore, already happening.

In this book, I will examine the various ways nanovision manifests, informs, and transforms the emerging culture of nanotechnology. We will quickly begin to see through the techno-deconstructive dynamic between blindness and insight, humanism and posthumanism, science and science fiction, that operates as the condition of possibility for engineering the future. In other words, we will begin to see through the singularity of nano-vision . . .

. . . and therein discover that the future is fully capable of accommodating not just one, but many, nanovisions.

1

NANOTECHNOLOGY IN

THE AGE OF POSTHUMAN ENGINEERING:

Science as Science Fiction

> Now nanotechnology had made nearly anything pos-
> sible, and so the cultural role in deciding what *should*
> be done with it had become far more important than
> imagining what *could* be done with it.
> —Neal Stephenson, *The Diamond Age*
>
> Long live the new flesh.
> —*Videodrome*

K. Eric Drexler, pioneer and popularizer of the emerging science of nanotechnology, has summarized the ultimate goal of this field as "thorough
and inexpensive control of the structure of matter."[1] Nanotechnology entails the practical manipulation of atoms; it is engineering conducted on
the molecular scale. Many scientists involved in this ambitious program
envision building nanoscopic machines, often called "assemblers" or "nanobots," that would be used to construct objects on an atom-by-atom basis.
Modeled largely on biological "machines" like enzymes, ribosomes, and
mitochondria—even the cell—these nanomachines would have specific
purposes, such as binding two chemical elements together or taking certain compounds apart, and would also be designed to replicate themselves
so that the speed and scale of molecular manufacturing may be increased.
Several different types of nanomachines would act together to build complex objects precise and reproducible down to every atomic variable. Other
researchers imagine using self-assembling macromolecular systems for

massively parallel data processing, leading to new computational capabilities more powerful than we can yet fathom. Still others suggest that with advances in nanoscale manipulation and visualization—currently exemplified by the field of scanning probe microscopy—we will eventually be able to program our material environments, placing individual atoms right where we want them with digital accuracy. In time, according to Mihail Roco, the senior advisor for nanotechnology at the National Science Foundation, we will "fundamentally control the properties and behavior of matter."[2] With its bold schemes to dominate materiality itself, nanotechnology has been prophesied to accomplish almost anything called forth by human desires.

THE TECHNOSCAPES AND DREAMSCAPES OF NANOTECHNOLOGY

Prophecies of the coming age of nanotechnology have run the gamut from the mundane to the fantastic: Designer nanoparticles will improve the performance of nearly all consumer products, from pharmaceuticals to shampoo. Smart nanofabrics will be woven into your carpet and clothing, programmed to constantly vaporize any dirt motes they encounter, keeping your house and your wardrobe perpetually clean. Nanomachines will be able to disassemble organic compounds, such as wood, oil, or sewage, and restructure the constituent carbon atoms into diamond crystals of predetermined size and shape for numerous purposes, including structural materials of unprecedented strength. Nanofactories will quickly and cheaply fabricate furniture, or car engines, or nutritious food, from a soup of appropriate elements. Nanobots will facilitate our exploration of space, synthesizing weightless lightsails to propel seamless spaceships throughout the universe. Nanosurgical devices will repair damaged human cells on the molecular level, healing injury, curing disease, prolonging life, perhaps annihilating death altogether.

Nanotechnology has been discussed extensively in these terms, and despite the fancifulness of certain nanoscenarios, it has become a robust and lucrative science whose cultural prominence has skyrocketed since the turn of the millennium. Many universities, laboratories, and companies around the world are now investigating nanotech possibilities, constituting a dense network—a technoscape—of individuals and institutions interested in the potential benefits of this nascent discipline. The U.S. National Nanotechnology Initiative (NNI), proposed by the Clinton administration in 2000 and augmented by the Bush administration in 2003 with the Twenty-first Century Nanotechnology Research and Development Act, offers funding and guidelines to promote nanotech breakthroughs. Other major nanotechnology initiatives have appeared in quick succession in the United Kingdom, Japan, the European Union, Switzerland, China, and elsewhere, coordinating thousands of international research sites. Extending the technoscape beyond its industrial and academic locations are foundations like the Center for Responsible Nanotechnology, the Institute for Nanotechnology, and, perhaps most prominently, the Foresight Institute, established in 1986 by Drexler and Christine Peterson. Hosting conferences, sponsoring publications and awards, the Foresight Institute (renamed the Foresight Nanotech Institute in 2005) has fashioned itself as a mecca of sorts in the multitude of nanotechnological endeavors spreading across the globe. This spread has been facilitated significantly by the Internet, whose role in nanoculture cannot be underestimated: while the sci.nanotech newsgroup may have been the only formalized venue for nanotheorizing online when it went live in 1988, the numerous websites, chat rooms, and blogs that disseminate daily information while also cultivating virtual social networks—including *Howard Lovy's Nanobot, Nanotechnology Now, Small Times*, and more—have lately become nearly uncountable. Since Drexler first proposed a potential program for research in 1986 with the publication of *Engines of Creation: The Coming Era of Nanotechnology*, nanotechnology has gained notoriety as a visionary science, and the technoscape has burgeoned.

Offering intellectual and commercial attractions, career opportunities, and a variety of research agendas, nanotechnology foresees a technocultural revolution that will, in a very short time, profoundly alter human life as we know it. The ability to perform molecular surgery on our bodies and our environments will have irrevocable social, economic, and epistemological effects; our relation to the world will change so utterly that even what it means to be human will seriously be challenged. But despite expanding interest in nanotech, despite proliferating ranks of researchers, despite international academic conferences, numerous doctoral dissertations, and thousands of publications, the promise of a world violently restructured by nanotechnology has yet to become reality.

Or has it?

Scientific journal articles reporting experimental achievements in nanotech, or reviewing the field, frequently speak of the technical advances still required for "the full potential of nanotechnology to be realized,"[3] of steps toward fulfilling the "dream of creating useful machines the size of a virus,"[4] of efforts that, if they "pan out, . . . could help researchers make everything from tiny pumps that release lifesaving drugs when needed to futuristic materials that heal themselves when damaged."[5] These texts—representative of the genre of popular and professional writing about nanotech that I will call "nanowriting"—incorporate individual experiments and accomplishments in nanoscience into a teleological narrative of "the evolution of nanotechnology,"[6] a progressivist account of a scientific field in which the climax, the "full potential," the "dream" of a nanotechnology capable of transforming garbage into gourmet meals and sending invisible surgeons through the bloodstream, is envisioned as *already inevitable.*

Nanowritings swathe their technical contents with the aura of destiny. The Nobel Prize–winning nanoscientist Richard Smalley has written that "there is a sense of inevitability that [future nanotech successes] will come in time," declaring that there "will come technologies that will be the best that they can ever be" and that "all manner of technologies will flow" from the current work of dedicated visionaries.[7] Christine Peterson, cofounder of the Foresight Institute, agrees that the "development of nanotechnology appears inevitable."[8] Likewise, Ray Kurzweil asserts that "nanotechnology is the inevitable end result of the ongoing miniaturization of technology of all kinds."[9] Hans Moravec concurs that "atomic scale construction is not just possible but inevitable in the foreseeable future. . . . Our accelerating technology will soon reach a kind of escape velocity that will carry us into a new and radically different world."[10] John G. Cramer, a theoretical physicist and science fiction novelist, writes that nanotech "ideas carry an 'air of inevitability' about them. The technology is coming."[11] Textual dispatches from the frontiers of nanoscience everywhere make such claims, drawing on the future as a known quantity, determined in advance. These publications freely and ubiquitously import the nanofuture into the research of today, rewriting the advances of tomorrow in the present tense.[12] Nanowritings speculate on scientific and technological discoveries that have not yet occurred, but they nonetheless deploy such fictionalized events to describe and to encourage preparation for the wide-scale consequences of what the nanotheorist B. C. Crandall describes as a "seemingly inevitable technological revolution."[13]

Even in one of the field's earliest articles to appear in a technical journal—an article that both proposed a new technology and inaugurated a new theoretical program—Drexler claimed that the incipient engineering science of molecular nanotechnology would have dramatic "implications for the present," as well as for "the long-range future of humanity."[14] Repeated throughout the technoscape, this narrative telos of nanotechnology—described as already given—is a vision of the "long-range future of humanity" utterly transfigured by present scientific developments. In other words, embedded within nanowriting is the implicit assumption that, even though the nanodreams have not yet come to fruition, nanotechnology has *already* changed the world.

Until recently, nanotheorists had yet to produce material counterparts to their adventurous mathematical models and computer simulations. Donald Eigler, a nanoscientist at IBM's Almaden Research Center, appraised the technical situation only a short time ago: "Nanotechnology is a vision, a hope to manufacture on the length scale of a few atoms." For the moment, he said, "nanotechnology doesn't exist."[15] Heedless of its existential status, though, nanotechnology as a research field has been strongly inclined to speculate on the far future and to prognosticate its role in the radical metamorphosis of human life. Which is why many skeptics and critics over the years have claimed that nanotechnology is less a science and more a science fiction. For instance, David E. H. Jones, a chemist at the University of Newcastle upon Tyne, once insinuated that nanotech is not a "realistic" science, and that, because its aspirations seem to violate certain natural limits of physics, "nanotechnology need not be taken seriously. It will remain just another exhibit in the freak-show that is the boundless-optimism school of technical forecasting."[16] Gary Stix, a staff writer for *Scientific American* and persistent critic of nanohype, has often compared Drexler's writings to the scientific romances of Jules Verne and H. G. Wells, suggesting that "real nanotechnology" is not to be found in these science fiction stories.[17] Furthermore, Stix maintains that nanowriting, a "subgenre of science fiction," damages the legitimacy of nanoscience in the public eye, and that "distinguishing between what's real and what's not" is essential for nanotech's prosperity.[18] Even George M. Whitesides, the Mallinckrodt Professor of Chemistry at Harvard University and a recognized authority in nanotechnology, has accused the entire field of being "an area prone to overblown promises, with speculation and nano-machines that are more likely found in *Star Trek* than in a laboratory."[19] For Whitesides, designs for self-replicating molecular machines in particular are "complete nonsense. . . . The level of hard science in these ideas is really very low."[20] Similarly, Steven M. Block, a biophysicist at Stanford University, has said that many nanoscientists, especially Drexler and the "cult of futurists" involved with the Foresight Institute, have been too influenced by the laughable expectations of science fiction and have gotten ahead of themselves; he proposes that for "real science to proceed, nanotechnologists ought to distance themselves from the giggle factor."[21]

Some critics have insisted that advanced atomic manipulation and engineering will not be physically possible for thermodynamic or quantum mechanical reasons; others have suggested that, without experimental verification to support its theories and imaginary miraculous devices, nanotechnology is not scientifically valid; many more have dismissed the long-range predictions made by nanowriting on the grounds that such speculation obscures the reality of present-day research and the appreciable accomplishments within the field. These critiques commonly and tactically oppose a vocabulary of "real science" to the term "science fiction," and, whether rejecting the entire field as mere fantasy or attempting to extricate the scientific facts of nanotech from their science fiction entanglements, charges of science fictionality have repeatedly called the epistemological status of nanotechnology into question.[22]

Nanotechnologists have responded with various rhetorical strategies intended to distance their science from the negative associations of science fiction. However, I argue that such strategies ultimately end up collapsing the distinction, reinforcing the science fiction aspects of nano at the same time as they rescue its scientific legitimacy. I hope to make clear that the scientific achievements of nanotechnology have been and will continue to be extraordinarily significant; but, without contradiction, nanotechnology is thoroughly science-fictional in imagining its own future, and the future of the world, as the product of scientific advances that have not yet occurred.[23]

Science fiction, in Darko Suvin's formalist account of the genre, is classically identified by the narratological deployment of a "novum"—a scientific or technological "cognitive innovation" as extrapolation or deviation from present-day realities—that becomes "'totalizing' in the sense that it [the novum] entails a change in the whole universe of the tale."[24] The diegesis of the science fiction story is an estranging "alternate reality logically necessitated by and proceeding from the narrative kernel of the novum."[25] Succinctly, science fiction assumes an element of transgression from contemporary scientific thought that in itself brings about the transformation of the world. It follows that nanowriting, positing the world turned upside down by the future advent of fully functional nanomachines, thereby falls into the domain of science fiction. Nanowriting performs radical ontological displacements within its texts and re-creates the world atom by atom as a crucial component of its extrapolative scientific method; but by employing this method, nanowriting becomes a postmodern genre that draws from, and contributes to, the fabulations of science fiction.[26] Science fiction is not a layer than can be stripped from nanoscience without loss, for it is the exclusive domain in which mature nanotechnology currently exists; it forms the horizon orienting the trajectory of much nanoscale research, and any eventual appearance of practical molecular manufacturing—transforming the world at a still unknown point in the future through a tremendous materialization of the fantastic—would remain marked with the semiotic residue of the science fiction novum. Accordingly, I suggest that nanotechnology should be viewed as simultaneously a science and a science fiction.

Jean Baudrillard has frequently written on the relationship of science to science fiction, contextualizing the dynamics of this relationship within his notion of hyperreality. Mapping schematically onto the "three orders of simulacra"[27]—the counterfeit, the reproduction, and the simulation—three orders of the speculative imaginary are described in his essay "Simulacra and Science Fiction": "To the first category [of simulacra] belongs the imagination of utopia. To the second corresponds science fiction, strictly speaking. To the third corresponds—is there an imaginary that might correspond to this order?" The question is open because the third-order imaginary is still in the process of becoming and is as yet unnamed. But within this imaginary, the boundary between the real and its representation deteriorates, and Baudrillard writes that, in the postmodern moment, "There is no real, there is no imaginary except at a certain distance. What happens when this distance, including that between the real and imaginary, tends to abolish itself, to be reabsorbed on behalf of the model?" The answer is the sedimentation of hyperreality, where the model becomes indistinguishable from the real, supplants the real, precedes the real, and finally is taken as more real than the real:

> The models no longer constitute either transcendence or projection, they no longer constitute the imaginary in relation to the real, they are themselves an anticipation of the real, and thus leave no room for any sort of fictional anticipation—they are immanent, and thus leave no room for any kind of imaginary transcendence. The field opened is that of simulation in the cybernetic sense, that is, of the manipulation of these models at every level (scenarios, the setting up of simulated situations, etc.) but then *nothing distinguishes this opera from the operation itself and the gestation of the real; there is no more fiction.*[28]

In the dichotomy of science versus science fiction, the advent of third-order simulacra or imaginaries announces that science and science fiction are no longer separable. The borderline between them is deconstructed. In the age of simulation, science and science fiction have become coterminous: "It is no longer possible to fabricate the unreal from the real, the imaginary from the givens of the real. The process will, rather, be the opposite: it will be to put decentered situations, models of simulation in place and to contrive to give them the feeling of the real, of the banal, of lived experience, to reinvent the real as fiction, precisely because it has disappeared from our life." At the moment when science emerges from within science fiction and we can no longer tell the difference, the real has retreated, and we are left only with the simulations of the hyperreal, where "there is neither fiction nor reality anymore," and "science fiction in this sense is no longer anywhere, and it is everywhere."[29]

The case of nanotechnology spectacularly illustrates the hyperreal disappearance of the divide between science and science fiction. The terminology of "real science" versus "science fiction" consistently used in the debates surrounding nanotech depends on the discursive logic of the real versus the simulacrum. Although each term may independently provide the illusion of having a positive referent—that is, "real science" might refer to a set of research and writing practices that adhere to and reveal facts of nature while being institutionally recognized as doing so, and "science fiction" might refer to a set of generically related fictional texts or writing practices that mimic such texts—when they are used to argue the cultural status of nanotechnology, real science and science fiction are nearly emptied of referential pretensions, becoming signifiers of unstable signifieds as they are forced into preestablished structural positions of "the real" and "the simulacrum." In this logic, science and science fiction negatively define each other, and though each is required for the other's symbolic existence, science fiction is the diminished and illegitimate term, the parasitical simulation of science.

To maintain that the categories of science and science fiction are supplemental constructs of each other is not to deny the material and political ramifications of discourse, for the fate of nanotechnology as a research field and the fates of real people working within it are strongly entwined with the language used. But I will show that the nanorhetoric mobilizing the logic of real science *opposed* to science fiction comes to undermine its own position, dissolving real science *into* science fiction and enacting the vanishing of the real, or the moment of hyperreal crisis when the real and "its" simulacrum appear as semiotic fabrications, when "the real" ("real science") can be demonstrated as simulation and "the simulation" ("science fiction") can be demonstrated as real, when dichotomies must be abandoned in favor of hybrids. Although the strict categories of real science and science fiction must be used to accomplish their deconstruction (or are deconstructed because of their use), they should be read as under erasure, for the relationship of science to science fiction is one not of dichotomy but of imbrication and symbiosis. Science fiction infuses science and vice versa, and vectors of influence point both ways. Inhabiting the liminal space traversed by these vectors are fields like nanotechnology that draw equally from the inscription practices of scientific research and science fiction narration, and only a more sutured concept—something like "science (fiction)"—adequately represents the technoscape of nanotechnology and its impact on the human future.

Nanotechnology is one particular example epitomizing the complex interface where science and science fiction bleed into each other.[30] Yet more significantly, nanotechnology is fully capable of engineering the future in its own hybrid image. Not only does the continued development of nanotechnology seemingly provide the means for making our material environments into the stuff of our wildest dreams, but nanotech's narratives of the "already inevitable" nanofuture ask us *even now* to reevaluate the foundations of our lived human realities and our expectations for the shape of things to come. Which is to say that the writing of nanotechnology, as much as or even more than any of its eagerly anticipated technological inventions, is already forging our conceptions of tomorrow. Unleashing its science fictions as science and thereby redrawing the contours of technoculture, nanotechnology instantiates the science-fictionizing of the world.

A cognitive shock, an epistemic virus, the science-fictionizing of socalled reality shatters conventional modes of thought and activates ways of seeing differently. As Donna Haraway has argued, this postmodern revelation that "the boundary between science fiction and social reality is an optical illusion" gives rise to a cyborg epistemology which threatens the matrix of humanism.[31] Similarly, Scott Bukatman sees the new subjectivity created by the science fictions of technoculture as a "terminal identity": "Terminal identity is a form of speech, as an essential cyborg formation, and a potentially subversive reconception of the subject that situates the human and the technological as coextensive, codependent, and mutually defining."[32] These cyborg fusions and science fiction technologies transfigure embodied experience, enabling the emergence of a posthuman subject that N. Katherine Hayles describes as "an amalgam, a collection of heterogeneous components, a material-informational entity whose boundaries undergo continuous construction and reconstruction."[33] I argue that nanotechnology is an active site of such cyborg boundary confusions and posthuman productivity, for within the technoscapes and dreamscapes of nanotechnology, the biological and the technological interpenetrate, science and science fiction merge, and our lives are rewritten by the imaginative gaze resulting from the splice—the new way of seeing that I have called nanovision. The possible parameters of human subjectivities and human bodies, the limits of somatic existence, are transformed by the invisible machinations of nanotechnology—both the nanowriting of today and the nanoengineering of the future—facilitating the eclipse of so-called man and the dawning of the posthuman condition.

NANOTECHNOLOGY AS SCIENCE, OR, THE NANORHETORIC

Nanotech is a vigorous scientific field anticipating a technological revolution of immense proportions in the near future, and K. Eric Drexler has long been at the vanguard of these anticipations (figure 5). Founder of the Foresight Institute as well as chief technical advisor for the molecular engineering company Nanorex, Drexler has impressive scientific credentials—a Ph.D. from MIT (1991), a former visiting appointment at Stanford (1986–91), a research fellowship at the Institute for Molecular Manufacturing (1991–2003), and numerous publications. His technical and popular writings have inspired many researchers to enter the field of nanoscience. So it might be surprising to recognize that Drexler's seminal *Engines of Creation*, outlining his program for nanotech research, is composed as a series of science-fictional vignettes. From spaceships to smart fabrics, from AI to immortality, *Engines of Creation* is a veritable checklist of science fiction clichés (Drexler's insistence on scientificity notwithstanding), and the book's narrative structure unfolds like a space opera: watch as brilliant nanoscientists seize control of the atom and lead humankind across the universe . . . and beyond!

This operatic excess of nanowriting—that genre of scientific text in which the already inevitable nanotech revolution can be glimpsed—characterizes even many of the technical publications by Drexler, Ralph Merkle, Wolfgang Heckl, Ted Sargent, Richard Smalley, Stuart Hameroff, Robert Freitas, Carlo Montemagno, J. Storrs Hall, and sundry other prophets of the nanofuture. Speculative and theoretical, the writings of these nanoscientists regularly demonstrate what is technologically possible but often not what has yet been accomplished, what has been successfully simulated but often not what has yet been realized. For example, Merkle writes that nanoscientists are working diligently to "transform nanotechnology from computer models into reality."[34] Likewise, Montemagno suggests that recent conceptual advances "make the idea of complex, molecular-sized engineered devices appear less like Science Fiction and more like an achievable goal."[35] Appearances notwithstanding, these texts frame their scientific arguments with vivid tales

5. K. Eric Drexler. A simulated diamondoid molecular bearing looms behind the visionary nanotechnologist. Photograph by Peter Menzel. © Peter Menzel/www.menzelphoto .com. Reproduced with permission.

of potential applications, which are firmly the stuff of the golden age of science fiction. Matter compilers, molecular surgeons, spaceships, space colonies, cryonics, autogenous robots, cyborgs, synthetic organisms, smart utility fogs, molecular cognition, extraterrestrial technological civilizations, and utopias abound in these publications, borrowing unabashedly from the repertoire of twentieth-century speculative literature.[36]

Consequently, the experimental evidence supporting the reality of nanotech has been marshaled to divide the science from its "sci-fi" associations. Nanotechnology is a realistic science, many researchers claim, because biological "nanomachines" like enzymes and viruses already exist in nature; there is no reason, then, why human engineers could not construct similar molecular devices.[37] Moreover, individual atoms can already be moved with relative ease using probe microscopes, and it seems certain that our technical abilities in this area will continue to improve. But even with nature as a model, and even with several highly publicized events in atomic manipulation, the tangible products of nanoresearch remain preliminary and exploratory. Which is why, according to Mihail Roco, nanotechnology was so widely perceived as pure science fiction in the years before the global wave of nanospeculation began to surge in 2000.[38] The infant field of nanoscience therefore advertised certain experimental results achieved during those early years as landmarks in the progress of the discipline toward becoming a "real science" (even despite the fact that many of the researchers involved in these landmarks may not have perceived themselves as "nanotechnologists" until well after the fact).[39] Some major proof-of-concept events claimed by nanoscience to defend against accusations of being nothing but science fiction have included the following:

— In 1981, Gerd Binnig and Heinrich Rohrer invented the scanning tunneling microscope (STM), which enabled real-space visualization and manipulation of individual atoms for the first time (figure 6).[40] Binnig and Rohrer were honored with the Nobel Prize in 1986, along with Ernst Ruska, the creator of the electron microscope. Also in 1986, Binnig, Christoph Gerber, and Calvin Quate developed the atomic force microscope (AFM), which, unlike the STM, can be used on nonconductive sample materials. Many other scanning probe microscopes soon joined the STM and the AFM, and this family of instruments now provides a variety of ways to interface with the nanoscale world.

— In 1987, Jean-Marie Lehn, Charles Pederson, and Donald Cram shared a Nobel Prize for their development of synthetic molecules with enzyme-like capabilities. Between 1987 and 1988, William DeGrado and his colleagues designed and created the world's first artificial protein.[41]

— In 1988, John Foster, Jane Frommer and Patrick Arnett at IBM successfully pinned organic molecules to a graphite surface and dissected them using an STM, suggesting the possibility of modifying or "editing" individual molecules.[42]

— In 1989, Donald Eigler and Erhard Schweizer maneuvered individual xenon atoms on a nickel surface with an STM, creating an iconic nanoscale image of the IBM logo (figure 7).[43]

— In 1996, Richard Smalley, Robert Curl, and Sir Harold Kroto received a Nobel Prize for their discovery of buckminsterfullerenes, a.k.a. "buckyballs" or C_{60}. By virtue of their remarkable strength, spherical form, and chemical properties, buckyballs can serve as robust structural materials, components of customized molecules, and containers for individual atoms. In 1991, Sumio Iijima discovered another type

6. Heinrich Rohrer (*left*) and Gerd Binnig (*right*) examine a first-generation scanning tunneling microscope at the IBM Zurich Research Laboratory (ca. 1981). Courtesy of IBM Zurich Research Laboratory.

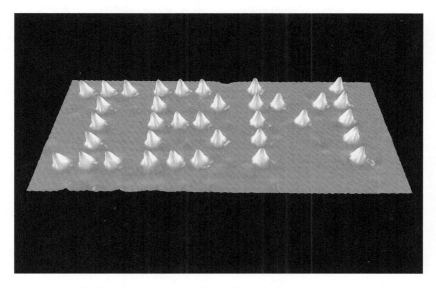

7. IBM logo made from xenon atoms (1989). Eigler and Schweizer's experiment in atomic writing quickly became one of the defining moments in the early history of nanotechnology. Courtesy of IBM Research, Almaden Research Center.

of fullerene: the nanotube. Nanotubes are durable cylinders of carbon that can be used for conductive or semiconductive electronic elements, probes or funnels for atomic positioning, "nanopencils" that deposit molecular ink, "nanotweezers" for moving atoms, and many other nanotech applications.[44]

— In the mid-1990s, nanoscientists created a range of spectacular nano-novelties, such as a "nanoabacus" (produced in 1996 by an IBM team led by James Gimzewski), a "nanotrain" (a large mobile molecule crawling along a molecular "track," synthesized by Viola Vogel), and several varieties of dynamic molecular motors.[45]

— In 1999, Hyojune Lee and Wilson Ho used an STM to bond an iron atom together with two carbon monoxide molecules, thereby mechanically assembling $Fe(CO)_2$ and demonstrating that molecular mechanosynthesis is experimentally feasible.[46]

Because these early technical accomplishments suggested *progression* toward the "full potential" of nanotech during the years when nano's reputation as a real science was most at stake, enthusiasts could maintain that the continued "evolution of nanotechnology" was a scientifically valid expectation. Since then, in the wake of the international nanotech frenzy that began around 2000, the ongoing production of nanoparticles and nanofilms, experiments with "quantum dots" (semiconducting nanocrystals), DNA computing, and the multiplication of research programs around the world have dramatically augmented the perception that the progression of nanotechnology is well under way. Even the questionable assimilation of research that once would simply have been called "chemistry" or "microphysics" under the umbrella term of "nanotechnology" has contributed to an increasing faith not only that nanotechnology has now become a real science but also that its inevitable advancement will fulfill our every dream.

Further evidence that nanotechnology is a real science, rather than a misguided fad, comes from its many signs of emergent disciplinarity, or even a synthetic transdisciplinarity.[47] That scientists from numerous research traditions are actively working and staking their reputations on the nanofuture might be evidence enough of a more or less common mission, and even the visible confrontation between various research programs and individual scientists (most notoriously Drexler and Smalley) seeking to shape the field can be seen as symptomatic of the tribulations of nanotechnology as a whole to attain the status of a unified scientific profession.[48] These agonistic struggles within the technoscape help to stabilize field-specific lexicons, as well as institutional structures supporting nanoresearch, and effectively establish conceptual boundaries within which various kinds of nanoscience are able to take place, both industrially and academically.[49]

Drexler taught an engineering course on molecular nanotechnology at Stanford University in 1989, and this early curricular inclusion supposedly indicated the emergent institutionalization of an already exciting field: "At Stanford, when I taught the first university course on nanotechnology, the room and hallway were packed on the first day, and the last entering student climbed through a window."[50] Such early, even premature, attention to pedagogy suggests an effort to plant nanotechnology in the ecology of academic disciplines from its first days. Indeed, although a textbook usually marks the trailing end or calcification of a scientific discipline rather than its evolutionary forefront, Drexler composed an advanced textbook on nanotech engineering and design called *Nanosystems: Molecular Machinery, Manufacturing, and Computation* (1992)—based on his doctoral dissertation—long before most scientists had even heard the word "nanotechnology."[51] Filled with the differential equations, quantum mechanical calculations, and structural diagrams that had been absent from his earlier publications, *Nanosystems* performed a certain legitimating function for what was still a maligned science, even in 1992. Dozens of other nano textbooks were soon published in succession, and Drexler's "first university course" has been followed by hundreds of international academic programs, institutes, courses, and workshops focusing on nano education—something Drexler thus seems to have foreseen already in 1989.

Since 1989, the Foresight Institute has sponsored annual conferences on nanotechnology, bringing in researchers from all over the world to define the goals, methods, and assumptions of the new technoscience. Governmental and university conferences, as well as online forums, have created spaces for professional networking and information exchange between the growing communities of self-fashioned nanotechnologists. There are even several scholarly journals, such as *Nanotechnology*, *Nature Nanotechnology*, and *Nano Letters*, that publish exclusively the most advanced research in the field. At present, all these approaches to fashioning a unified nanotechnology discipline or a transdisciplinary "trading zone" have managed to achieve only a chaotic multidisciplinary muddle whose different constituents often do not see eye to eye.[52] But the failure of disciplinary cohesion so far has certainly not diminished the sense that as a scientific field—perhaps unified in name only, but still a field—nanotechnology has *arrived*. The first nanotech start-up company, Zyvex, appeared in Richardson, Texas, in 1997, intending to develop nanodevices like Drexler's assembler in less than a decade. Zyvex has been followed by a boom of nanocommerce in Silicon Valley and other regions where industrial speculation and venture capital abundantly flow. Many of the actors involved in nanotech see these and other signs pointing to a grand "convergence" of numerous sciences and industries at the nano-scale.[53] Across academia, industry, and politics, nanotechnology seems here to stay. "Finally," as the nanobusiness analyst Josh Wolfe puts it, "we are living in a nanotech age."[54]

So nano certainly *looks* like real science — could there be any doubt? — and the people promoting the field have long been trying really hard to show why it is nothing like science fiction, despite the fantastic nature of its many futurological predictions. Again and again, boosters tell us that nanotech-nology is not, *absolutely not*, science fiction. This fact is insisted on, every-where and loudly. For example, Smalley writes:

> Real nanotechnology is more amazing than any pipe dream. It is closing in on structural materials stronger than anything we've known; on com-puters the size of molecules; on complete diagnostic laboratories smaller than your thumbnail; on ways to painlessly cook cancer cells to death; on buildings that stay up despite storms, earthquakes and attacks.
>
> Set pulp fiction aside. The genuine nanocosm has sci-fi beat six ways to Sunday.[55]

Nanotechnology will be awesome, it seems, and it beats up "sci-fi" roundly, reducing it to "pulp." The main argument enforcing this battle line emerges, again, from the logic of the real versus the simulacrum. Specifically, nano-scientists insist that their visions of the future are "genuine," grounded in "real science," while the "pipe dream" futures described in science fiction are not. Take Drexler's comments on science fiction in *Engines of Creation*:

> By now, most readers will have noted that this [nanotechnology] . . . sounds like science fiction. Some may be pleased, some dismayed that future possibilities do in fact have this quality. Some, though, may feel that "sounding like science fiction" is somehow grounds for dismissal. This feeling is common and deserves scrutiny.
>
> Technology and science fiction have long shared a curious relation-ship. In imagining future technologies, SF writers have been guided partly by science, partly by human longings, and partly by the market demand for bizarre stories. Some of their imaginings later become real,

because ideas that seem plausible and interesting in fiction sometimes prove possible and attractive in actuality. What is more, when scientists and engineers foresee a dramatic possibility, such as rocket-powered spaceflight, SF writers commonly grab the idea and popularize it.

Later, when engineering advances bring these possibilities closer to realization, other writers examine the facts and describe the prospects. These descriptions, unless they are quite abstract, then sound like science fiction. Future possibilities will often resemble today's fiction, just as robots, spaceships, and computers resemble yesterday's fiction. How could it be otherwise? Dramatic new technologies sound like science fiction because science fiction authors, despite their frequent fantasies, aren't blind and have a professional interest in the area.

Science fiction authors often fictionalize (that is, counterfeit) the scientific content of their stories to "explain" dramatic technical advances, lump them together with this bogus science, and ignore the lot. This is unfortunate. When engineers project future abilities, they test their ideas, evolving them to fit our best understanding of the laws of nature. The resulting concepts must be distinguished from ideas evolved to fit the demands of paperback fiction. Our lives will depend upon it. (92–93)

I quote this passage at length because of its several remarkable qualities intended to rescue nanotechnology from the ghetto of science fiction. While the first paragraph begins the radical task of reconciling science and science fiction, juxtaposing the languages of "possibility" and "fact," Drexler quickly departs from this goal and instead firmly separates science, and particularly nanotechnology, from the "fantasies" of fiction. He clarifies the assumed directional flow of reality into fiction: when science fiction is "real," the writer has either landed on reality by chance or "grabbed" the idea from science. Drexler thus distinguishes science fiction writers from "other writers" and "engineers" who "examine the facts" (presumably Drexler fits into this category). He employs the idea of the "counterfeit" to describe science fiction as a mimetic representation similar to, but ontologically distinct from, reality. He divides "our best understanding of the laws of nature" (Drexler's writing) from "the demands of paperback fiction" (science fiction), concluding that because of the dangerously real consequences made possible by nanotech, our very lives depend on maintaining this division. What further rationale for recognizing the barrier between science and science fiction could one need?

Thus Drexler seemingly secures his work as science, but another tactic deployed by defenders of nanotech is to exclude Drexler and his sympathizers from the technoscape entirely. This strategy acknowledges and foregrounds the intractable science fictionisms of Drexler's science and thereby pronounces him a pariah, in effect preserving the rest of nanotech as "real science."[56] For example, Don Eigler (of the xenon IBM logo) has audaciously declared that Drexler "has had no influence on what goes on in nanoscience. Based on what little I've seen, Drexler's ideas are nanofanciful notions that are not very meaningful."[57] Mark Reed, a nanoelectronics researcher and Harold Hodgkinson Professor of Engineering and Applied Science at Yale University, has said, "There has been no experimental verification for any of Drexler's ideas. We're now starting to do the *real* measurements and demonstrations at that scale to get a *realistic* view of what can be fabricated and how things work. It's time for the *real* nanotech to stand up."[58] The force of this argument comes from the deluge of the "real," which, repeated ad nauseam, appears to drown Drexler and friends and condemns them to the irrationalities of their nanodreams. Again we see the rhetorical establishment of a powerful dichotomy of science versus science fiction, this time constructed within the technoscape itself.

A final tactic used by nanorhetoricians, both Drexlerians and Drexler detractors, is the oft-repeated story about the genesis of nanotech. I will call this foundational narrative the "Feynman origin myth." The story goes (and it is rehearsed by legions of researchers in the field, posted on their web pages, and retold in their publications) that on December 29, 1959, Richard Feynman delivered a talk entitled "There's Plenty of Room at the Bottom" to the American Physical Society at the California Institute of Technology.[59] Here Feynman suggested the possibility of engineering on the molecular level, arguing that the "principles of physics, as far as I can see, do not speak against the possibility of maneuvering things atom by atom. It is not an attempt to violate any laws; it is something, in principle, that can be done."[60] Feynman further asserted that something like nanotech is "a development which I think cannot be avoided." Quotations and paraphrases of his statements run rampant throughout the discourse network as ammunition in the ongoing war to legitimate nanotechnology.[61] Such recourse to Feynman's speech has given rise to the belief that he originated, authorized, and established nanotechnology. Assertions like "This possibility [of nanotechnology] was first advanced by Richard Feynman in 1959" and "Richard Feynman originated the idea of nanotechnology, or molecular machines, in the early 1960s" are commonplace and have taken on the status of truisms.[62] Feynman's talk is continually invoked to prove that nanotechnology is a real science, but not because of the talk's theoretical, mathematical, or experimental sophistication; indeed, judging from the language used—the numerous appearances of "possibility," "in principle," "I think," and the telling "it would be, in principle, possible (I think)"—it is clear that the talk was just as speculative as (if not more than) any article penned by Drexler, Merkle, or their associates.

The Feynman origin myth is resurrected over and over again as an easy way of garnering scientific authority. How better to ensure that your science is valid than to have one of the most famous physicists of all time pronouncing on the "possibility" of your field? It is not uncommon for nanorhetoricians, when referencing the talk, to remind their audience that Feynman won the 1965 Nobel Prize in physics. "And when a Nobel Prize winner says something," a recent nanotechnology primer reminds us, "you listen . . . and try to understand."[63] Merkle candidly reveals that name recognition and cultural capital are the main values of this tactic when he writes: "One of the arguments in favor of nanotechnology is that Richard Feynman, in a remarkable talk given in 1959, said that, 'The principles of physics, as far as I can see, do not speak against the possibility of maneuvering things atom by atom.'"[64] The argument is clearly not *what* Feynman said but that *he* said it. The argument hinges on his unique vision, what he "can see," something special about Feynman's scientific ability that transforms a speculative statement into a description of reality (figure 8).[65] A frank example of fetishizing the author and the origin (the Foresight Institute even offers a Feynman Prize), Feynman's talk grounds nanotechnology not in the real but in authoritative discourse. Nevertheless, the Feynman origin myth is perceived as dissociating nanotechnology from science fiction.

To its credit, nanotech has been amazingly successful in vindicating itself as real science, as something very different from science fiction despite how much it may seem like science fiction. The anti-sf rhetoric has even made nanodreams appear more like inevitabilities to a larger audience. From 1992, when Drexler first unveiled a wonderful nanofuture to the U.S. government, to the implementation of the 2001 National Nanotechnology Initiative, the foundations for which grew out of congressional testimonies by Smalley, Merkle, and other key figures in the field, nanorhetoric triumphed in transforming the visions of science fiction into manifest and lucrative national ventures.[66] Even President Clinton, announcing the National Nano-

8. Richard Feynman speaks of microscopic machines. Captured from a video recording of Feynman's 1984 speech "Tiny Machines," a redux of "There's Plenty of Room at the Bottom" (published on DVD by Sound Photosynthesis as "Tiny Machines: The Feynman Lecture on Nanotechnology"). Here Feynman bodies forth the incredible smallness of imaginary molecular devices with a pinch of his fingers—a digital gesture typical of nano-discourse. © 2004 Sound Photosynthesis.

technology Initiative at Caltech on January 21, 2000, demonstrated his absorption of nanorhetoric by citing the 1959 Feynman talk, along with a few imaginary coming attractions of the nanofuture, as evidence for the decisive role that nanotechnology will play in bringing about an "era of unparalleled promise."[67] Thus, despite its many skeptics and determined critics, nanotech has managed to secure its professional future by combining fantastic speculation with concerted attacks on science fiction. Indeed, considering nanotech's rapid expansion in academia and industry, the reputable scientists involved, and its current high profile, there appears little doubt that nanotech is real science.

However, the "sci-fi" anxieties that haunt the defenders of nanotechnology disclose its scandalous proximity to science fiction, and, I argue, only rhetoric is maintaining the separation. Furthermore, this rhetoric thoroughly deconstructs itself in a futile struggle for boundary articulation that has already been lost.

NANOTECHNOLOGY AS SCIENCE FICTION,
OR, DECONSTRUCTING THE NANORHETORIC

Recall Drexler's arguments regarding science fiction. Drexler must explicitly distinguish his science from paperback fiction because his nanonarratives borrow extensively from preexisting genre conventions. Drexler's stories — like those found throughout nanowriting — describe the world transformed by imagined feats of science and engineering relegated to the unspecified future; and even when denying the science fictionality of his vignettes by emphasizing that they are "scientifically sound," he cannot avoid drawing attention to the fact that they do, after all, "sound like science fiction." Although Drexler confirms the conventional assumption that science is the real, and science fiction its imaginary simulacrum, when he says that his science "sounds *like*" fiction, he reverses this assumed order. Science fiction has anticipated science, and the ensuing science is not ultimately delineated from science fiction by Drexler's arguments.

Although Drexler distinguishes science fiction writing from his kind of writing through the criterion of mimesis, science fiction writers who "grab the idea [from science] and popularize it" are not logically different from writers who "examine the facts" of science and popularize them, as *Engines of Creation* is intended to do. Along the same lines, the criterion that Drexler's stories are scientifically sound while science fiction stories are (presumably) not is challenged when he acknowledges that science fiction "imaginings" frequently "become real" (again reversing the presumed order). Science and science fiction dynamically and frequently shift structural positions in Drexler's writing, both suggested to be inhabited by "the real" at the same time as each paradoxically appears to simulate the other. That is to say, the real has become simulation, and the simulation has become real.

None of these inconsistencies mean that Drexler is not writing good science; they do mean that the boundary between writers of science fiction and writers of what Drexler calls "theoretical applied science," like himself, is hopelessly blurred. Tellingly, Drexler has himself forayed into the production of genre science fiction texts, writing an introduction to the short story collection *Nanodreams* (1995), where he discusses the importance of science fiction in assessing future technologies.[68] The unavoidable failure of the dichotomy between science and science fiction occurs when Drexler, having apparently given up the endeavor, also calls the scenarios described in *Engines of Creation* "science fiction dreams."[69]

Thus the division between writers of science fiction and writers of "theoretical applied science" or "exploratory engineering" is destabilized and confused. "Scientifically sound," according to Drexler, can be a quality of both kinds of writing—destroying the criterion, erasing the division. Ultimately, Drexler's nanowriting indicates that science fiction precedes and supersedes "its" science, echoing Baudrillard's "precession of simulacra": the simulacra coming before, displacing and supplanting, making the real seem to be the not-real, the science to be science fiction.[70]

Determining that Drexler's version of nanotechnology is inseparable from its science fictionisms would apparently make the tactic of excluding him from the field more effective. After all, if his writing is indeed science fiction, then he is not, according to Reed, part of "the real nanotech." However, attempts to banish Drexler from the field he helped establish actually have the ironic effect of highlighting the science fictionality of nano. When Eigler states that Drexler "has had no influence on what goes on in nanoscience," he is disregarding Drexler's seminal technical publications and the considerable contributions of the Foresight Institute. (His appraisal also seems rather infelicitous, considering that he and his colleagues would later cite Drexler's *Nanosystems* as background to their own experiments with computational molecule cascades.)[71] Furthermore, Eigler's comments flatly contradict the vast expanses of the technoscape that recognize Drexler's influence[72]—including Smalley, who once said that Drexler "has had tremendous effect on the field through his books,"[73] and, despite his conviction that Drexlerian self-replicating assemblers are "not possible," he credited Drexler with motivating his own illustrious career in nano: "Reading [*Engines of Creation*] was the trigger event that started my own journey in nanotechnology."[74] When Reed says that Drexler's ideas have not been experimentally verified and therefore are not part of the "real" nanotech, he is disregarding the validity of all theoretical science—clearly a problematic move. And even if Drexler could be fashioned as a heretical character on the fringes of the technoscape, it remains the case that other scientists perhaps more sociologically central to "real nanotechnology" employ with regularity the same science fiction tropes in their nanowritings as Drexler does in his. In terms of their common reliance on speculative narrational modes to bridge the gaps between current technoscientific reality and future promises, there is no difference between them.[75] Consequently, Drexler cannot be so simply exiled. He has convinced not only individual nanoscientists but also governmental funding boards about the inevitable nanofuture.[76] The same "science fiction dreams" that inform Drexler's writing permeate the nanotechnological imagination everywhere, right to the very center of "real science." Accordingly, nanotechnology needs to acknowledge the heavy speculation that remains fundamental for its own development as a research field. After all, having proclaimed that Drexler is "science fiction" and "not real," yet ultimately obliged to recognize his influence, this tactic for expelling science fiction from science backfires on itself.

Even Merkle's response to these exclusionary efforts eventually backfires. In a letter to the editor of *Technology Review*, he writes:

> While I am happy to see the increasing interest in nanotechnology, I was disappointed by your special report on this important subject. Mark Reed summarized one common thread of the articles when he said, "There has been no experimental verification for any of (Eric) Drexler's ideas." Presumably this includes the proposal to use self-replication to reduce manufacturing costs. The fact that the planet is covered by self-replicating systems is at odds with Reed's claim.
>
> Self-replicating programmable molecular manufacturing systems, a.k.a. assemblers, are not living systems. This difference lets Reed argue that they have never before been built and their feasibility has not been experimentally verified. Of course, this statement applies to anything we have not built. Reed has discovered the universal criticism. Proposals for a lunar landing in 1960? Heavier-than-air flight before the Wright brothers? Babbage's proposal to build a computer before 1850? No experimental verification. Case closed.[77]

Merkle musters a "fact" (that self-replicating systems abound in nature) in support of Drexler and builds an argument for the validity of scientific speculation, countering Reed's implication that Drexler's science is not "real." Drexler is salvaged, put back on the secure ground of reality. But while accomplishing Drexler's reassimilation into the field, Merkle also winds up equating nanotechnology with science fiction. Merkle suggests that nanotechnology is a real science, even though it lacks experimental verification, because proposals for a lunar landing in 1960, considerations of heavier-than-air flight before the Wright brothers, and Babbage's idea for a computer had no experimental verification and yet these ideas eventually found verification after time. "Case closed," he writes. But, of course, speculations about a moon voyage, heavier-than-air flight, and computers of various sorts had existed long before their "real" incarnations—think of the stories of Jules Verne, H. G. Wells, Hugo Gernsback, Isaac Asimov, Robert A. Heinlein, Arthur C. Clarke, and countless others—all of which were and still are clearly marked as science fiction. Thus, in recuperating the speculations of nanowriting, Merkle solidifies the relay between nanotechnology and science fiction. Before moon voyages, air flight, and computers there was science fiction; before the nanotechnology revolution of the future there is the anticipatory nanotechnology of today. Nanotechnology is science fiction. Case closed?

The dissolving boundary between science and science fiction in nanowriting elsewhere occurs as intertextuality, in the sense that loci of meaning within nanowritings frequently depend on a larger web of texts, both science and science fiction, that enable their signification. In this respect, nanowritings are what Jonathan Culler describes as "intertextual constructs" that "can be read only in relation to other texts, and [are] made possible by the codes which animate the discursive spaces of a culture."[78] For example, the concept of the "Diamond Age"—describing how the nanotechnology era will be historicized relative to the Stone Age, the Bronze Age, the Silicon Age, and so on—appears formatively in science fiction, such as Neal Stephenson's nanotech novel *The Diamond Age*, as well as in subsequent nanowritings like Gregory Benford's "A Scientist's Notebook: A Diamond Age," Merkle's survey article, "It's a Small, Small, Small, Small World," and Edward Reitman's textbook, *Molecular Engineering of Nanosystems*. Each text simply assumes the reader's familiarity with the terminology deployed by the others.

Stephenson's novel, similarly, describes a "Merkle Hall" located within the nanotech corporation Design Works, whose ceiling, reminiscent of Michelangelo's Sistine Chapel, is covered with a fresco depicting the pantheon of nanotech, wherein Feynman, Merkle, and Drexler mingle with more fictional personalities, all "reposed on a numinous buckyball."[79] Fact and fiction merge in the blender of nanowriting, where allusions are creatively drawn from both technical reports and popular novels.

Stephenson's nod to the "numinous buckyball" draws Richard Smalley and his fullerene research into the realm of science fiction. But to some degree, this intertextual move has been anticipated by Smalley's own nanowriting. After all, despite his many disparaging remarks about science fiction over the years, Smalley has occasionally referenced the "space elevator" from Arthur C. Clarke's *The Fountains of Paradise* as a potential achievement for fullerene nanotechnology. In an early account of the boundless functionality of nanotubes, Smalley and Boris Yakobson write: "In a 1978 science-fiction novel called *Fountains of Paradise* Arthur Clarke described a strong filament or cable being lowered from a geosynchronous satellite and used by the engineers of the future to move things up and down from earth—a space elevator. . . . None of the materials now known to humankind get close to such strength. Fullerene cables someday may."[80] Literary fancies here irrupt into scientific writing, luring nanotechnology to the stars. Who, then, are the real "engineers of the future"?

Or how about this: narratives in the *Star Trek* franchise are now rife with the basic premises of nanotech, featured, for instance, in the episode "Evolution" (1989) from *Star Trek: The Next Generation,* or in the Borg systems depicted in episodes such as "Scorpion" (1997) from *Star Trek: Voyager,* and "Regeneration" (2003) from *Star Trek: Enterprise.*[81] Nano even takes center stage in Steven Piziks's *Star Trek: Voyager* novel *The Nanotech War* (2002). But no less are some technical nanowritings now rife with the basic premises of *Star Trek.* For example, in their scientific article "Star Trek Replicators and Diatom Nanotechnology" (2003), the researchers Ryan W. Drum and Richard Gordon describe the usage of diatoms for nanoscale engineering:

> The most remarkable invention in the new field of diatom nanotechnology is that akin to the replicator in the science fiction television and movie series *Star Trek.* Diatom shells are placed in an atmosphere of magnesium gas at 900°C for 4 hr. Apparently an atom for atom substitution of magnesium for silicon occurs with no change in 3D shape. Thus silicon oxide, a material of limited usefulness in nanotechnology, is transformed into magnesium oxide. . . . [Such gas/solid reactions] should "replicate" diatom silica shells into a variety of oxides. The list of uses envisaged includes microcapsules for medications, sensors, optical diffraction gratings and actuators, or templates for any of these.[82]

Textual allusion here functions constitutively, in that an explicitly fictive reference transfigures the object of nanotech research. The chemical reaction morphs into a "remarkable invention" akin to the *Star Trek* "replicator"—a device for creating desired objects from a feed of raw materials—whose overdetermined role as novum now suggests a whole range of speculative applications stretching into the future. The mission of the nano laboratory: to boldly go where no man has gone before!

The constitutive role of science fiction allusion arises even more strikingly in J. Storrs Hall's theoretical elaboration of a nanotech "Utility Fog"— a pervasive substance for complete environmental control and universal human-machine interface.[83] The Fog would consist of a swarm of nanomachinic "foglets" dispersed in the air, designed to change properties and simulate any range of normal materials. In presenting exploratory engineering designs for this technology, Hall's essay "Utility Fog: The Stuff That Dreams Are Made Of" (1996) relies on a diffusion of science fiction tropes and witty references to many canonical science fiction texts, including *Forbidden Planet* (1956), Heinlein's "The Roads Must Roll" (1940), Verne's *From the Earth to the Moon* (1865), Wells's *The Shape of Things to Come* (1933), and Čapek's *R.U.R* (1920), suggesting that nanotechnological thinking is essentially a process of writing from the margins of other fictional futures, other textual worlds.

In a later account, Hall notes that the Utility Fog's capability to simulate nearly anything, producing material objects straight out of the nano-saturated air, will thwart standard notions of the "real" and the "virtual": "One thing in a Fog world that would be more difficult than ours would be telling what was real and what wasn't. . . . Utility Fog mixes virtual and real. . . . I'm sure that as we gain experience with the partly virtual world that is coming, we'll invent new words and concepts necessary to deal with it."[84] But this hyperreality of the coming nanofuture already informs Hall's writing of the present, for it seems that designs for Utility Fog originated from mixing real science with superhero fantasy: "I'll . . . explain where 'Utility Fog' came from. First the stuff fills the air like fog, and you walk around in it. It would look a lot like fog as well. As for the 'Utility' part, remember the *Batman* TV series from the 1960s? Whenever Batman needed some gadget, lo and behold, there it was in his 'utility belt.' There seemed to be no end of what the belt could produce—and it was always right at hand."[85] Within nanowriting, the facile permeability of these worlds of science and fiction, the ease with which concepts and signs traffic between them, challenges any stringent boundrification.

This semiotic drift is perhaps most visible in the case of professional scientists contributing to the field of nanowriting who are also themselves professional science fiction writers. The electrical engineer Bart Kosko, the microbiologist Joan Slonczewski, the aerospace engineer Wil McCarthy, the physicists Gregory Benford, David Brin, and John G. Cramer, and other literati spanning both speculative and real science have helped shape the technoscapes of nanotechnology, whether through their own nanoscale experiments, or their involvement in organizations like the Foresight Institute and the Center for Responsible Nanotechnology, or their prominent pronouncements on the coming nanofuture.[86] At the same time, they have helped shape the dreamscapes of nanotechnology by writing richly imagined novels dramatizing many of the same issues explored by their nonfiction nanowritings.[87] Which comes first, the science or the fiction? They seem, rather, to engender each other. According to Slonczewski, her tales of nanobiotechnology and microbial intelligence feed from her microbiology research, and vice versa: "Microbiology is getting stranger than anything in science fiction; when my stories include the latest research, reviewers call it 'impossible.' Nevertheless, writing science fiction is good practice for grant proposals."[88] Estranging fictions inspire stranger grant proposals, seemingly "impossible" experiments inspire even-stranger stories, leading to more outlandish grant proposals, and so on: a positive cybernetic loop with accelerating returns.

The utility of this kind of cognitive estrangement for financing scientific projects and generating public support has also been clear to certain actors involved in governmental nano initiatives. William Sims Bainbridge, for one, has served as an officer of the U.S. National Science Foundation since 1992, playing a key part in federal nanoscience legislation and development of the NNI. A professional sociologist, Bainbridge has also written expertly on the ideological dimensions of science fiction and the genre's ability to "promote spaceflight and the exploratory mind."[89] In his studies of public attitudes toward nanotechnology, he has analyzed the messages of popular nanofiction novels and designed questionnaires featuring his own fictitious "vignettes."[90] He has even suggested the proactive use of science fiction for engineering new techno-religious movements to advance the social progress of science.[91] That his reports on the effects of nanoconvergence—including human-nanomachine interfaces, nanoengineered spaceships, interplanetary civilizations, and high-performance nanobots—are themselves awash in wild scenarios appears completely consistent with his literary agenda for sculpting technoculture.[92] For it would seem that anticipatory narratives help to incubate the nanofuture:

> Our civilization will benefit from artistic creativity that is inspired by nanotechnology. Science fiction can also inspire young people to enter technical fields, and it can communicate to a wide audience the excitement that real scientists and engineers experience in their work. We should applaud the individual writers who have begun promulgating visionary images of the future of the field, and we cannot expect them to limit their imaginations to the proven nanoscale techniques of today. . . . Nanotechnology can gain support for science more broadly by attracting people's uninhibited hopes. Call it fantasy, if you will, but fantasy is one of humanity's greatest sources of innovation. . . . There is no telling

which of the wildest ideas of science fiction might turn out to be surprisingly feasible, or which ideas will inspire young people to become scientists and engineers, building the future of humanity.[93]

The "excitement that real scientists and engineers experience in their work" infects a wider public via science-fictional vectors. The "real" scientific sense of wonder in technical labor would therefore appear to be commensurate with, or even identical to, the emotional affect transmitted by science fiction—understood here as delight in the work of technoscience—which improves our whole civilization. For Bainbridge, the future is made not by limiting the imagination to techniques of today but by unleashing it as an emotive tool for inspiring young people, advancing technical knowledge and labor, and fostering public support for research. Again, the tactics of separating nanotech from the science fiction with which it is complicit would seem to be self-defeating on nearly every level.

As a final bit of evidence, let's return to the Feynman origin myth. Despite nanorhetoricians' frequent citations of the talk to support the realness of their discipline, the talk itself sits awkwardly with such a purpose. We have seen the indeterminacy and speculative nature of the language Feynman uses, and strikingly, the talk is composed as a series of science fiction stories, just like Drexler's *Engines of Creation*. Feynman tells stories about tiny writing, tiny computers, the actual visualization of an atom, human surgery accomplished by "swallow[ing] the surgeon," and "completely automatic factories"—certainly not impossibilities, but nonetheless the conceits of numerous genre science fiction narratives long before Feynman stepped to the podium. Thoroughly penetrated by the literary imagination, it is no coincidence that Feynman's nanotech looks just like Drexler's nanotech, fabricated from the same "science fiction dreams."

The Feynman origin myth thus contains in itself the deconstruction of the nanotech–science fiction dichotomy. The cavalier way in which the myth is used both by Drexlerians and by those who challenge Drexler's vision is a further indication of its self-deconstructive tendencies. Consider, for example, the response of Thomas N. Theis (of the IBM Research Division) to the *Technology Review* article where Reed implies that Drexler's nanotech is not real: "Congratulations on your review. . . . Your writers clearly distinguished hype from hard science and vision from reality. I was reminded of Richard Feynman's famous 1959 after-dinner talk. . . . Feynman managed to foreshadow decades of advances. . . . I know that his vision influenced at least a few of the individuals who have made these [hard science] things happen."[94] That Theis can speak of "vision" opposed to "reality" in one sentence, and of Feynman's "vision" that *contributed* to hard (i.e., real) science in another, reveals the ease of appropriating such a myth for one's own purposes, the impossibility of simply excluding Drexler's "vision" from the field, and the blurring of science and science fiction within the Feynman talk. After all, if vision is opposed to reality, then Feynman's talk abandoned reality entirely.

Even as a genesis story, the Feynman myth only succeeds in making science fiction of nanotechnology. Nanotechnology's status as real science is supposedly strengthened because it was founded and authorized by the great Richard Feynman. But this origin is not an origin, and its displacement unravels its legacy. The Feynman myth would work only if it clearly had no precedents, if it was truly an "original" event in intellectual history, if Feynman had offered a unique, programmatic conception of how nanotechnology was to be accomplished. Yet such is not the case: Feynman merely depicted a speculative vision of a possible technology; and science fiction writers, as they have done with so many things, had already beaten him there. Technologies, theoretical concepts, and thought experiments identifiably similar to those circulating through current visions of nanotechnology appear in numerous stories from the first half of the twentieth century. For example, Ray Cummings's *The Girl in the Golden Atom* (1923; based on his short story of 1919) features the invention of a subatomic microscope, which eventually leads to the physical exploration of romantic worlds inside an atom. Theodore Sturgeon's "Microcosmic God" (1941) represents artificially evolved microentities called "Neoterics," whose small size enables them to engineer eutactically and create incredible new devices. Robert A. Heinlein's "Waldo" (1942) discovers a top-down method for interacting with infinitesimal materials. Eric Frank Russell's "Hobbyist" (1947) describes a mysterious factory that manufactures living organisms through atom-by-atom assembly, located on a planet that may be God's own workshop. Hal Clement's *Needle* (1949) imagines an alien being made up entirely of viroid particles who diffuses into a young boy's body, enhancing his physical capabilities and acting as an internal surgeon. James Blish's "Surface Tension" (1952) dramatizes the construction of miniature technologies inside an aqueous lifeworld whose features are dominated by intermolecular forces. And finally, Philip K. Dick's "Autofac" (1954) concerns fully "automatic factories" sustained by self-replicating microscopic machinery. All these stories were published well before Feynman gave his now-mythical talk.

Although it is unknown whether Feynman personally read any of these science fiction stories, his friend Albert R. Hibbs (senior staff scientist at the Jet Propulsion Laboratory) did read "Waldo" and described it to Feynman in the period just before Feynman composed his talk.[95] And indeed, Heinlein's influence haunts Feynman's depiction of nanotechnology. In Heinlein's novella, the eponymous genius Waldo has invented devices — known as "waldoes" — which are mechanical hands of varying sizes, slaved to a set of master hands attached to a human operator. Heinlein writes that the "secondary waldoes, whose actions could be controlled by Waldo himself by means of his primaries," are used to make smaller and smaller copies of themselves ("[Waldo] used the tiny waldoes to create tinier ones"), ultimately permitting Waldo to directly manipulate microscopic materials by means of his own human hands.[96] Heinlein thus hypothesizes a method for molecular engineering that Feynman in his talk, without crediting his source, offers as a means to "arrange the atoms one by one the way we want them." Feynman describes his proposed system:

> [It would be based on] a set of master and slave hands, so that by operating a set of levers here, you control the "hands" there. . . . I want to build . . . a master-slave system which operates electrically. But I want the slaves to be made especially carefully by modern large-scale machinists so that they are one-fourth the scale of the "hands" that you ordinarily maneuver. So you have a scheme by which you can do things at one-quarter scale anyway — the little servo motors with little hands play with little nuts and bolts; they drill little holes; they are four times smaller. Aha! So I manufacture [with these hands] . . . still another set of hands again relatively one-quarter size! . . . Thus I can now manipulate the one-sixteenth size hands. Well, you get the principle from there on.[97]

The originality of the Feynman myth crumbles, for we can see that Feynman's talk emerges from genre science fiction. Feynman's method of molecular manipulation is borrowed from Heinlein. Even the proposition for internal microscopic surgery — a notion Feynman credits to Albert Hibbs — was already proclaimed as an "original" idea by Heinlein in the "Waldo" novella. Heinlein writes that microscopic surgery via microscopic machines "had *never been seen before*, but Waldo gave that aspect little thought; no one had told him that such surgery was unheard-of."[98] The mythologized order of precedence is therefore reversed, for it becomes evident that speculations of nanotech were freely circulating in the discourse of science fiction long before science "grabbed the idea." If we really want to locate an "origin" to nanotechnology, it is not to Feynman that we must look, but to science fiction.

Consequently, I reiterate that in the case of nanotech we have a situation where simulation has preceded and enveloped "real" science, where the line between science and science fiction is blurred, made porous, and effaced. It even seems likely that this hybridity has been responsible for nanotech's recent financial success; companies have been founded and government officials have been awed less by nanotech's real accomplishments than by its dream of the future, its promise of a world reborn: its science fiction indistinguishable from its science. Rapidly becoming a major actor in the science-fictionizing of technoculture—along with certain other interstitial sciences and technologies, such as virtual reality, cybernetics, cloning, exobiology, astronautics, artificial intelligence, and artificial life—nanotechnology exerts strong symbolic influence over the way we conceptualize the world and ourselves. In other words, as a science (fiction) with enormous cultural resources and increasing historical significance, nanotechnology claims for itself a powerful role in the human future and the future of the human.

POSTHUMAN ENGINEERING

The birth of nanotechnology provokes the hyperreal collapse of humanistic discourse, puncturing the fragile membrane between real and simulation, science and science fiction, organism and machine. With its technoscapes and dreamscapes irreducibly interlaminated, nano heralds metamorphic futures and cyborganic discontinuities. For in both its speculative-theoretical and applied-engineering modes, nanotechnology unbuilds those constructions of human thought, as well as those forms of human embodiment, based on the security of presence and stability—terrorizing presentist humanism from the vantage point of an already inevitable future. If the discovery of leakages across ruptured conceptual boundaries denatures the domain of humanism, entreating us to "pass beyond man and humanism, the name of man being the name of that being who, throughout the history of metaphysics or of ontotheology—in other words, throughout his entire history—has dreamed of full presence, the reassuring foundation, the origin and the end of play,"[99] then the arrival of nano and its hyperreal way of seeing performs exactly this kind of crucial destabilization within the human present. For nano puts all foundations in play even at the atomic level; it

upsets presentism and the security of linear time; it erases epistemic divisions even as it multiplies ontic possibilities. Nano torques the edges of the
human imaginary, marking a site of humanism's blindness—the shadow of
an onrushing Singularity.

The singular breakdown of humanism accelerates through increasing col-
lisions between human flesh and technology, where the interfaces mediate
the emergence of new posthuman spaces, hybrid realities of the machinic,
the virtual, and the meaty.[100] Where bodies bleed with machineries, where
science bleeds with science fiction, the secure enveloping tissues of the
human subject—cognitive, corporeal, and otherwise—rip apart. Within these
wounds, these traumatic crash sites that become ever more refined through
technical reductions approaching the quantum limits of fabrication, the
natural and the constructed, the human and the nonhuman, wash together
in a molecular flow. This confluence and convergence at the nanoscale thus
make possible a radical reshaping of reality, atom by atom—a reshaping
of reality that, while still a fiction, is no less already a fact. Even right now,
the science (fiction) of nanotechnology enacts the techno-deconstruction of
humanism, forcing us to think otherwise through its narratives of corporeal
reconfiguration from beyond the temporal horizon, fabricating new fields
of embodiment and facilitating our becoming posthuman by envisioning a
future where the world and the body have been made into the stuff of science fiction dreams.

Posthuman narratives of bodily ambiguation, fragmenting and exceeding
the bodies we know, restructure our somatic experiences. As Kelly Hurley
has argued, the textual operations of posthuman narratives work to "disallow
human specificity on every level, to evacuate the 'human subject' in terms
of bodily, species, sexual, and psychological identity," generating instead
discursive zones for self-alienation and self-refashioning.[101] By radically re-
vising the body, such narratives offer alternate modes of identification to
what is merely standardized as human. Toward this end, nanotechnology
everywhere produces images of bodily ambiguation and articulates an alternative way of seeing—the subversive technoscientific gaze of nanovision—
in myriad future-shock stories circulating within the technoscape and beyond. Whether deployed in the form of novels, films, or technical scientific
reports, these posthuman nanonarratives directly impact and modify our
present.[102] Many nanoscientists seem to confirm the immediate and tangible effects of such stories in their conviction that humanity has already
been remade by a colossal technological revolution that may so far have
occurred only in fiction yet necessitates ever more vigilant "foresight." That

several key figures in nano also openly participate in transhumanist and extropian movements, endorsing the redesign of human biology through advanced molecular manufacturing and other still-fictive technologies, powerfully suggests the affective force of visionary nanonarratives.[103]

Whether utopian dreams or catastrophic nightmares, nanonarratives resist traditional humanist interpretations by repeatedly depicting the future in terms that disequilibrate the human body. From the eroticized collective consciousness of the Drummers in Stephenson's *The Diamond Age* (1995), to the lycanthropic transformations of Dean Koontz's *Midnight* (1989), to the permeability of "enlivened" city structures and body structures in Kathleen Ann Goonan's *Queen City Jazz* (1994), to the metamorphosis of the entire human population into billowing sheets of sentient brown sludge in Greg Bear's *Blood Music* (1985), posthuman bodies in nanonarratives are never stable, never idealized, never normative, never confined; the limits of posthuman corporeality are as wide as the nanovisual imagination. Nanovision disrupts the configurations of the human body, rebuilding the body without commitment to the forms given by nature or culture. Nanovision is an active instrument of posthuman engineering.

Rather than purveying a posthumanism in which the subject is in danger of losing the body, nanovision sees posthuman subjectivities resulting from embodied transformations.[104] Embodiment is fundamental to nanovision because, for the science of nano, *matter* profoundly *matters*. Nanovision respects no unitary construct above the atom, reducing everything to a broadly programmable materiality and demolishing metaphysical categories of identity. Accordingly, it does not support any sort of abstracted, theoretical construction of the body because it unbounds the body, putting its surfaces and interiors into constant flux. Posthuman bodies conditioned by nanovision are therefore always individuated experiences of embodiment in an endless array of possible bodily conformations, where all skins and membranes are fair game.

Nanovision entails a cyborg logic, imploding the separation between the biological and the technological. As we have seen, one of the arguments legitimating nanotechnology is that biological machines like ribosomes and enzymes and cells are real, and consequently there is nothing impossible about engineering such nanomachines. But the very ease of describing biological objects as machines indicates the cyborgism of nanotech, its logic of prosthesis, its construction of bodies and machines as mutually constitutive. According to Carlo Montemagno: "Within all living organisms [exist] the original nanomachines. . . . Life is the result of a large number of carefully orchestrated processes that are facilitated by a complex consortium of elegant nanoscopic machines."[105] Nanotechnology envisions the components of the living body and mechanical objects as indistinguishable and subsequently uses the biological machine *as the model* for the nanomachine, achieving a terminal circularity. Nanovision removes all intellectual boundaries between organism and technology—as Drexler puts it, nanovision causes "the distinction between hardware and life . . . to blur"[106]—and human bodies become posthuman cyborgs, inextricably entwined, interpenetrant, and merged with the mechanical nanodevices *already inside them.*

Having become cyborganic machines, bodies in the grasp of nanovision can be reassembled or reproduced with engineering specificity. Unlike genomic cloning, which merely provides genotypic but not necessarily phenotypic identity, the copying fidelity of nanotechnology is so exact that copies would have precise identity down to the atomic level. Feynman (following Heinlein) foresaw this in his talk: "All of our devices can be mass produced so that they are absolutely perfect copies of one another."[107] The ability of nanodevices to produce exact copies—copies of themselves, copies of their constructions—is fundamental to nanovision, and it is perhaps not entirely a coincidence that for more than a decade Merkle directed the groundbreaking Computational Nanotechnology Project for Xerox.[108] The potential for nanotechnology to reproduce anything exactly, accurate in every atomic detail, or to reconstruct anything into an identical copy of anything else, leads to posthuman nanonarratives that, undermining our conceptions of identity and origin(ality), need not become literalized to have transformed the architectures of our somatic experience. Posthuman narratives ask us to envision otherwise, thereby opening up new possibilities of corporeality that change the way we conceive ourselves. Such possibilities are illustrated by the following series of nanoscenarios:

— A wooden chair, reprogrammed on the molecular level, can be transformed into a diamond table, its woody "chairness" subtly and efficiently morphed into crystalline "tableness." Nanovision undermines essentialism, insisting that every "thing" is simply a temporary arrangement of atoms that can be endlessly restructured. Becoming overcomes being.

— A wooden chair can be transformed into a living fish. There is no magic here, merely a precise rearrangement of molecules. Life arises instantly from dead material; as Drexler writes, nanovision reveals that "nature draws no line between living and nonliving."[109]

— A fish can be transformed into a human (i.e., *Homo sapiens*). The resulting human could even be a specific person like Sigourney Weaver (posthuman icon from the *Alien* films), identical to the movie star in every respect: DNA, proteins, phospholipids, neurotransmitters, memories.

— A human, subjected to nanomachines carrying the data set for another human, can suddenly become someone else, and back again. Human Alpha and Human Beta share the same matter, they occupy the same space; although they have different identities, although they are different people, they are smeared together across time as *phases* of each other.

— A woman can be metamorphosed into a man, or vice versa, or in various partial combinations. Mono-, inter-, and transsexuality can be manifested in a single figure. Tissues, hormones, and chromosomes can be refabricated. The posthuman body is thus queered: sex and sexuality made infinitely malleable, sexual difference slipping into sexual indeterminacy or deferral.

— A human body can become the copy of an already existing human body. Say, for example, Harrison Ford (posthuman icon from *Blade Runner*) transforms into Sigourney Weaver. Then there are two Sigourneys, identical down to the memories, even down to the belief that each is Sigourney Weaver and the other is the copy. There is no possible way of telling them apart, no possible way of telling which was the "original." Someone might ask, "Will the real Sigourney please stand up," but inevitably they both will. More disturbing than clones or android replicants, which merely mimic, these nanocopies actually *are*. Nanovision again destroys the difference between real and simulacrum.

— Nanotechnology can devise a matter transporter to facilitate human travel across great distances of space.[110] At one end, nanobots dismantle the human traveler atom by atom, recording the location of each molecule, until the traveler is just a pile of disorganized material. The nanobots feed data into a computer system, which instructs another group of nanobots at the terminal end of the transporter, working from a feed of appropriate elements, to reassemble the human traveler exactly as he or she had been at the proximal end. The traveler will have no memory of the trip but will emerge precisely as he or she was when the process began; though made from different atoms, the traveler is still the same person. Embodiment has been distributed across a spatial divide and between separate accumulations of matter. Furthermore, the data can be reused to construct multiple, identical copies of the traveler. Personhood can be duplicated, flesh xeroxed, minds mimeographed.

— Human bodies can be modified well beyond the confines of experience, becoming alien formations or improbable mélanges. Nanotechnology empowers posthuman imaginations to achieve outlandish physical alterations. (How many tentacles would you like to have?)

— Finally, nanovision enables us to think beyond human boundaries in a tragic sense, for nanotechnology can also bring about a posthuman future where all humanity has ceased to exist and nothing new emerges from the wreckage. This fate is made possible by insidious nanoweapons of mass destruction, or the nanocalyptic hypothesis of out-of-control nanobots turning the entire biosphere into "gray goo."[111] While providing a means to engineer new posthuman embodiment, nanotech also provides a means to engineer posthuman extinction.

As these scenarios suggest, nanotechnology has unprecedented effects on the way we are able to intuit our bodies, our biologies, our subjectivities, our technologies, and the world we share with other organisms. Whether positing the liberation of human potential or the annihilation of organic life on this planet, nanovision demands that we think outside the realms of the human and humanism. Nanovision makes our bodies cyborg and redefines our material experiences, redraws our cognitive borders, and reimagines our future. Accordingly, even before the full potential of a working nanotechnology has been realized, we have already become posthuman. Indeed, posthuman subjects abound in nanoliterature, and although science fiction novels like Ian McDonald's *Necroville* (1994), James L. Halperin's *The First Immortal* (1998), and Michael Flynn's *The Nanotech Chronicles* (1991) imagine posthuman nanomodified bodies as appearing at some ambiguous point in the future, other "nonfiction" posthuman beings exist already, right now, within the popular and professional writings of nanoscientists. As real, embodied, material entities, enmeshed in the semiosis of nanovision, these posthumans are found at nanotechnology's intersection with cryonics.

Drexler, Merkle, and other nanoscientists are deeply involved in the idea of freezing and preserving human bodies, or parts of human bodies, until the proper nanotechnology has been developed to revive and heal them. Freeze the body now, and eventually nanotechnology will resurrect the person, reversing not only the cellular damage caused by the freezing process, but also the damage that originally caused the person to die, maybe even building an entirely new body for the cryonaut. Cryonic science is not simply tangentially related to nanotechnology but has become a principal extension of nanovision—evidenced by the ubiquitous discussions of cryonics at all levels of nanodiscourse, from fanzines to university conferences.[112] Furthermore, Merkle is a former director of the Alcor Life Extension Foundation, a cryonics institute founded in 1972, and he also hosts a cryonics web page; Kosko and Drexler have served on the scientific advisory board of the Alcor Foundation and have written extensively about cryonics in their books and scientific journal articles.[113]

Even in Drexler's first nanotech publication, cryonic resuscitation is evoked when he writes that the "eventual development of the ability [of nanotechnology] to repair freezing damage [to cells] (and to circumvent cold damage during thawing) has consequences for the preservation of biological materials today, provided a sufficiently long-range perspective is taken."[114] Drexler thus implies that projected technologies of the future determine how we should deal with human tissues and human bodies in the present. Again nanowriting uses the language of the "already inevitable" and assumes that the full potential of nanotech has essentially been realized, temporal distance notwithstanding. Consequently, as deployed within the discourse of nanotechnology, the fact that cryonic techniques are currently in use means that nanomodified bodies are among us even now. Those who are dead but cryonically frozen have been encoded by nanovision as already revived, as already outside the humanistic dichotomy of dead or alive, as already voyagers into a brave new world of nanotech splendor . . . as already posthuman.

This nanovisionary encoding of the cryonaut is evident when Drexler writes of cryonic resurrection in the science fiction present tense, collapsing present and future, medical reality and technological fantasy, human death and posthuman revivification, into a single proleptic episode of *Engines of Creation*. Drexler tells of a hypothetical contemporary patient who "has expired because of a heart attack. . . . The patient is soon placed in biostasis to prevent irreversible dissolution. . . . Years pass. . . . [During this time, physicians learn to] use cell repair technology to resuscitate patients in biostasis. . . . Cell repair machines are pumped through the blood vessels [of the patient] and enter the cells. Repairs commence. . . . At last, the sleeper wakes refreshed to the light of a new day—and to the sight of old friends."[115] By way of alluding to H. G. Wells's *When the Sleeper Wakes* (1899), a canonic science fiction depiction of sleeping into the future, Drexler validates and necessitates present-day acts of cryonic freezing within his prophecy of the coming nano era. While indicative of nanowriting's dependence on the conventions of genre science fiction, this passage more significantly indicates how nanowriting's implosion of science into science fiction transmutes formerly human subjects into posthuman entities, amalgams of discourse and corporeality, biology and technology. Drexler's cryonaut becomes posthuman at the moment of being incorporated into nanonarrative, thereby surviving human death and becoming reborn through cyborg interpenetration with nanomachines. And though the cryonaut in Drexler's story is purely hypothetical, other more specific cryonauts are made posthuman through the same mangle of nanovision.

Take, for example, Walt Disney—perhaps the world's most famous cryonically preserved character. According to urban legend, Disney was frozen immediately upon his death in 1966 and placed in cold storage at an unnamed cryonics institute (though some rumors have gone so far as to suggest that Disney's icy tomb actually lies beneath the "Pirates of the Caribbean" ride at Disneyland).[116] In a wonderful semiotic tangle, the discourses of nanotechnology, cryonics, hyperreality, and posthumanism all converge under the sign of Disney. The viral expansion of Disneyism, the "Disneyfication" of postmodern culture, renders society itself hyperreal: the legend of Walt's own cryonic suspension is a telling symptom.[117] As we might expect, then, nanoscientists tell us that the miasma of hyperreality belching forth from the many Disney factories around the world will be enhanced dramatically by the advent of nanotechnology. For example, the nanoscientist and aerospace engineer Tom McKendree suggests that the "simulations" at Disneyland and other heightened-reality theme parks will become even more of "a total experience" through nanotech's ability to "make the fantasies real."[118] As if in anticipation, Epcot at Walt Disney World has already hosted two separate interactive nanotechnology exhibits at its "Innoventions" attraction—"It's a Nano World" (2004) and "Too Small to See: Zoom into Nanotechnology" (2006–7)—where visitors are surrounded by nanoscale structures and invited to play with molecules.[119] Projecting a futureworld of "hands-on" molecular manipulation, these exhibitions pedagogically reproduce the same logic that structured the old Disneyland ride "Adventure through Inner Space" (1967–85), which simulated shrinking its visitors with a "Mighty Microscope," ferrying them on a fantastic voyage inside an atom. While "Adventure through Inner Space" was once marked as science fantasy, it now emerges as a kind of retroactive prophecy.

Disneyism thus seems to be recapitulated and boosted by the imagineering capabilities of nanotechnology. To be sure, in Heinlein's "Waldo," the nanotech laboratory feels like a Disneyland simulation: "The ubiquitous waldoes, the insubstantial quality of the furniture, and the casual use of all the walls as work or storage surfaces, gave the place a madly fantastic air. [The visitor] felt as if he were caught in a Disney" (38–39). Or maybe it is Disneyland that now feels like Waldo's laboratory: Cory Doctorow's *Down and Out in the Magic Kingdom* (2003), set in a future where the dead can be revived inside "nano-based canopic jars" and the past can be preserved inside Disney resorts, depicts a refurbished Haunted Mansion whose animatronic undead are puppeteered by "human operators—telecontrollers, working with waldoes" (204, 85). With similar uncanny insight, Paul J. McAuley's novel of the nanofuture, *Fairyland* (1995), narrates the ascendancy of nanotech-enhanced nonhuman entities from within the Magic Kingdom outside Paris. McAuley describes the end of man beginning "in the early hours of the morning after the fall of the Magic Kingdom. . . . As the humans retreat into their dreams, brave new creatures will claim the world" (359). The rise and fall of the Magic Kingdom, allegorizing our hyperreal condition and our "retreat into dreams," thus marks the prelude to the posthuman nanotechnology era. So we begin to see why Disney "the man" materializes at the point where nano merges with cryonics.

Consider Merkle's "It's a Small, Small, Small, Small World" article: the title evokes the small world of atoms and assemblers purveyed by nanotechnology and, simultaneously, the "It's a Small World" ride at Disneyland and other Disney resorts, whose infectious and repetitious song ("It's a small world, after all! It's a small, small world!") metonymically stands for the Disneyscape as a whole. Disneyism is thus imported into nanowriting as a metaphor for the nanoworld itself, and appropriately so—for not only does this figural resonance reveal the embeddedness of nanovision in the plane of hyperreality, where science and science fiction are one and the same, but furthermore, Walt's crystallized body would thereby melt into the Tomorrowland-like nanofuture that enables his return from the dead. Merkle details the coming "Diamond Age" of nanotechnology, when the "ability to build molecule by molecule could also give us surgical instruments of such precision that they could operate on the cells and even the molecules from which we are made," and as many nanowriters have explained, such surgical precision will surely bring about cryonic resurrection.[120] Although Disney may be on ice, waiting to be reborn through the advances of nanotechnology, within nanowriting—where a "small world" of quotidian miracles is deemed already accomplished, where "nanotechnology will inevitably appear regardless of what we do or don't do"[121]—Disney the sleeper already wakes. The future is now, and through the textual machinations of nanowriting that permit preserved human bodies to surmount their own deaths, Walt Disney himself has been transmuted into a posthuman creature of flesh, machines, and hypersigns.

If nanovision's symbolic reprocessing of cryonauts like Walt Disney is any indication, then the transformation of the world envisioned by nanowriting is highly performative, and posthuman evolution has already begun. Accordingly, if nanotech is turning us posthuman, a critical scrutiny of the direction that nanotechnology takes and an engaged involvement in the corresponding changes to our lives and bodies are required to ensure that becoming posthuman is accomplished on our own terms. In *The Diamond Age*, Stephenson issues a note of caution as his novel replicates the narrative of nanotech inevitability, writing that "nanotechnology had made nearly anything possible, and so the cultural role in deciding what *should* be done with it had become far more important than imagining what *could* be done with it."[122] Nanotechnology empowers us to write our own posthuman future, but considering the massive biological, ecological, corporeal, and cultural changes heralded by nanovision (be they utopic or apocalyptic), as voyagers into the future, we must exercise the necessary foresight.

Indeed, foresight is a note that echoes throughout the technoscapes and dreamscapes of nanotechnology, from popular novels to experimental reports, as both a warning and an enticement. Nanotechnology and all its implications are on the horizon, bodied forth by the speculations of science and of fiction. With the nanofuture in sight, we must prepare for our posthuman condition . . . for it may be a small world, after all.

2

SMALL WORLDS:

Beyond the Limits of Fabrication

The only way to discover the limits of the possible is to
go beyond them into the impossible.
—Arthur C. Clarke, *Profiles of the Future*

Welcome to the desert of the real.
—*The Matrix*

"It's a small world." This old adage has always expressed the uncanny experience of discovering the familiar where only the unfamiliar had been expected, the known amid the unknown, the homely and the domestic within
the space of alienation—the expression of sudden *unheimliche* surprise that
Freud understood to recall, or call back, "something which is familiar and
old-established in the mind and which has become alienated."[1] But when
appropriated for the theme song of a multinational conglomerate like the
Walt Disney Company, the trope of the "small world" becomes instead an expression of the *impossibility* of uncanny experience in the postmodern world.
With the global expansion of mediated signals and simulations, the televisual deployment of reality as hyperreality, the world, as Marshall McLuhan has shown us, condenses into a local media landscape: "As electrically
contracted, the globe is no more than a village."[2] Wrapped within the cir-
culations of electronic data and transnational capital, the open space once
available for a sensational encounter with the foreign now vanishes from the
face of the earth. As Fredric Jameson writes, "The new space that emerges
involves the suppression of distance . . . and the relentless saturation of
any remaining voids and empty places, to the point where the postmodern

body . . . is now exposed to a perceptual barrage of immediacy from which all sheltering layers and intervening mediations have been removed."[3] The cutaneous adhesion of the contractile world allows everything to be felt at once: everything is now familiar and present. So rather than standing as a trace of the uncanny, of the affective jolt of finding something from one world to inhabit another, the phrase "It's a small world, after all," now comes only to insist that there is no more uncanny, simply because there is virtually no space which has not already been domesticated.

The widespread appeal of this Disneyfied slogan within the discourse on nanotechnology lies not simply in the fact that nanoscience deals with a very small world—a world of atoms, electrons, and macromolecular systems—but also in the suggestion that nanoscience may be radically reducing the lived dimensions of our own human world. Nanowritings from across science and industry affirm that a global shrinkage is at hand: "Nanotechnology makes a small world even smaller."[4] Heretofore, the full implications of Disney's vision for a truly diminutive world have been unfathomable: "When Walt Disney's characters introduced their song, *It's a Small World after All*, none of us, by any stretch of our imagination, conceived how small the world might become someday. . . . Nanotechnology in the new Millennium will produce various devices that will be SMALL, Small and smaller. . . . [This] new technology will change nearly every man-made object in the first 100 years of the Millennium."[5] The furthest "stretch of our imagination," a speculative gesture outward toward the future, as if an expansion, finds its limit—the point of its failure—at the dawn of the nanotechnology millennium, which portends transformations beyond what we can conceive in making the world smaller than we ever thought possible. But where the imagination fails, the future takes place; the imagination can now only play catch-up, because the future has already condensed into the present. Anticipation contracts into technological immediacy. As Terry Michalske, founding director of the Center for Integrated Nanotechnologies at Sandia and Los Alamos National Laboratories, has said: "We can really do things we can't imagine right now."[6]

Deliriously, this contraction of the space and time of anticipation, the onrushing of an unimaginable future into the present with a velocity greater than our ability to prepare for it—the discovery of a Singularity conterminous with the technoscapes of nanotechnology—would seem to be heralded by Disney's "It's a Small World, after All": "Everyone is familiar with this Disney theme song, which tends to run through your head for days after hearing it—often despite your efforts to turn it off! But the reality of small-world technology has arrived."[7] We don't have to wait for an "after all" for the world to shrink, because nanotechnology is here now, and this "small-world technology" has already brought the future into "reality." It *is* a small world, *after* all: a temporal confusion or tense interweaving of the future with the now, a tension at the surface of immediacy that suggests we have already surpassed a threshold in historical time: "We are entering a new age of technology in this millennium. Its impact will be as significant as when mankind progressed from the stone age to the iron age."[8]

After *all*. The end of an era—certainly the end of the human era—or even the end of time as such. The millenarianism evident within these pronouncements of a smaller and smaller world produced in the wake of nanotechnology—the eschatological small world, after all—and its rapturous sense that "small-world technology," in fabricating the world from the bottom up, discloses a profoundly *new* world beyond the already concluded present, thus takes as its gospel the self-replicating meme of hyper-real utopianism, a musical word-virus launched from within the "Happiest Place on Earth," a piece of infectious media now perceived to encode a prefiguration or a prophecy of nanotechnology, as if to prove that once again and even before it was imagined, even beyond the furthest stretch of the imagination back in 1964 "when Walt Disney's characters introduced their song," nanotechnology was already inevitable: "But if nanotechnology, the science of the tiny, grows (shrinks?) into its potential, everything from ultra light cars, aircraft, and spacecraft made out of diamond, to ultra fast diamond computers, could result. . . . Did you leave Disney's theme park's 'It's a Small World' and incessantly hum its theme song, 'It's a small world after all. . . .' for the rest of the day? I wonder if Disney ever imagined just how prescient his song might become."[9] The prescience of seeing in advance that, after all, the world will be small gives the arrival of nanoscience all the aura of destiny. Walt Disney and Richard Feynman are thus joined as links in the signifying chains of nanotechnology, imagineered as early prophets of a "small-world technology" that would, in "growing (shrinking)" into its potential, in collapsing or conflating science fiction with lived reality, bring the unimaginable into being.

NANO . . . WORLD . . . PICTURE

This prescience beyond the furthest stretch of the imagination, this foresight for the otherwise unthinkable new millennium manufactured by small-world technology, this unique ability to presage the future which is regularly attributed to Feynman, Drexler, and Disney (a future for which Walt can now be understood, at least mythically, to have prepared his own frozen body decades in advance), depends on seeing small. "I have seen the future, and it is small," writes the physicist Larry Smarr.[10] Similarly, the quantum nanoscientist Gerard Milburn has written, "The world of the quantum may be bizarre, but it is our world and our future."[11] We see such assertions everywhere in nanodiscourse: "Deep underground, in a microscope the size of a truck, Stanford scientists have seen the future. And it is very, very small."[12] These statements manifest a rhetorical impression of the future into the nanoscale and the quantum, a sense that the future itself can be seen within the unfamiliar territory of atoms and molecules, and vice versa. This collapse of worlds—the discovery of "our world" and "our future" in the nanoscale present—is a direct effect of nanovision. Nanovision describes a simultaneity of the specular and the speculative, the imaged and the imagined, the scientific and the science-fictional: a concatenation of the nanovisual ("it's a small world") and the nanovisionary ("after all").

I am going to claim that the propensity of nanoscience to see small accounts for its prophetic destining of the future—that is, its belief that it can see the future—and also its logic of control, its structuring ethos that "the emerging nanotechnology may provide humanity with unprecedented control over the material world," that it will enable a "shaping of the world atom by atom,"[13] or even a "reshaping of everyday life"[14] through its mastery of every atomic bit of quotidian reality. I aim to show that this aspiration for mastery is both produced by, and enabled through, novel scopic practices. The technologies of visualization that have helped inaugurate nanoresearch—specifically, probe microscopes such as the scanning tunneling microscope, or STM (figure 9)—produce an image of the so-called nanoworld that provides scientists with the sense of having mastered the molecular, bearing witness to the already inevitable transformation of the world, a future that has already arrived.

Nanovision, the scopic logic of nanoscience, does not originate with the technical apparatus. It goes back at least to a handful of science fiction stories from earlier in the twentieth century that anticipate the rise of nanoscale engineering—a few of which I will discuss in detail shortly. But nanovision is nevertheless reinforced and modulated by its technical instantiation in instruments like the STM. The narratives, metaphors, and textual figurations emergent from nanovision that both predate and surround the STM as a discursive field influence not only the way that nanorhetoric has used the tunneling microscope in its ongoing efforts to establish the "reality"

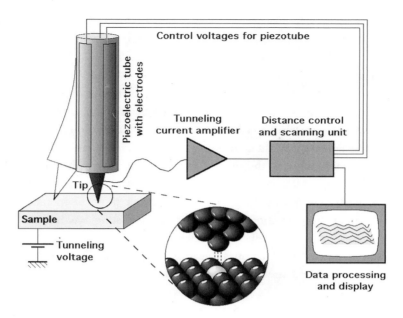

9. Schematic diagram of the scanning tunneling microscope (STM) by Michael Schmid, Technische Universität Wien, from the IAP/TU Wien STM Gallery. Reproduced by permission of Michael Schmid under the Creative Commons Attribution ShareAlike 2.0 Austria License, 2006.

and the "novelty" of nanotechnology, but also the way that researchers engage with the tunneling microscope at the level of physical embodiment. In other words, the embeddedness of the experimental instrumentation of nanoscience within the tropology characteristic of nanovision—whose features we began to see in chapter 1—establishes a feedback loop between the discursive field and the hardware. If the STM functions as the material incarnation of nanovision—that is, one possible instrument that enables nanovision to take place—it does so only because the tropology has already conditioned the instrument to be perceived as working in a way commensurate with established nanorhetoric.

This tropological conditioning of the instrument and its operation constitutes what I would call its *tropic protocol*: the diagrammatic and symbolic dimension underwriting and orienting the signifying operations of probe microscopy, the tropical horizon toward which these processes veer and against which they turn, the semiotic algorithm that sets out in advance an available vocabulary and an available phenomenology for the dynamic interaction of a human user with the instrument. Peter Galison has written: "Instruments embody—literally—powerful currents emanating from cultures far beyond the shores of a master equation or an ontological hypothesis. . . . The machines sit squarely in the laboratory and yet always link the laboratory to other places and practices."[15] The tropic protocol would be one of the specific linking passages or correspondences between the instrument, the larger ocean of nanodiscourse, and the many interrelated imaging practices of nanotechnology described all together by the term "nanovision."[16] While it determines neither all possible meanings nor all possible ways of seeing available to users of the STM, the chain of interlinked tropes anchoring the epistemic frame on which the STM rests nevertheless produces certain discernible effects.[17] Specifically, I will show how a tropic protocol operating through the imaging productions of the STM gives rise to the "small world" effect of nanotechnology. Or rather, this small-world effect might more precisely be designated as *affect*, because the ability to picture the nanoscale— an ability that transforms the world, shrinks it, and collapses the future into the present—emerges within STM-user assemblages as a raw intensity, rendered as a sensation of proprioceptive control.

The ability to "see" atoms within an optical register, the technical capability to survey the molecular terrain as if with one's own sensory apparatus, appears to inspire a drive toward complete domination of the material world. In 1990 at the IBM Almaden Research Center in San Jose, California, Donald Eigler and Erhard Schweizer introduced one of the most famous images of nanotechnology in action: the IBM logo inscribed on a nickel plate with xenon atoms (figure 7). This image was produced in the course of experiments conducted with an STM to study the adsorption and ordering of xenon on metal surfaces. In its imaging mode, the STM rasters a fine metallic tip across an electrically conducting sample, keeping the tip only a few nanometers from the surface of the sample. When an electric current is applied, electrons leap across the empty space between sample and tip through a phenomenon known as "quantum tunneling." In keeping this tunneling current constant, variations in the height of the tip as it scans laterally correspond to variations in surface topography. This topographic data can be rendered into a digital image of the nanoscale details of the surface, as well as any atoms that may be on top of it (called "adatoms"). While taking such nonoptical "pictures" of xenon adatoms, Eigler and Schweizer saw the future of atomic mastery unfold from within the digital image.[18]

The scientists quickly observed that by increasing the tunneling current, they could grab adatoms with the STM tip and drag them along the surface; they could "push, pull, pick up and put down surface atoms using the tip of the microscope."[19] The STM image allowed them "to locate the atom to be moved and to target its destination"—that is, to see both the present and the future of the atom's location—and this destination, this destiny of the xenon atoms under the scrutiny of IBM scientists, was the symbolic structure of the corporate logo.[20] The enframing of the nanoscale surface by the image— the IBM logo produced by the STM, manufactured in real time according to the tropic protocol of the symbolic order, that is, of the "logo" as such— makes future possibilities present, "evident" to our eyes: "This capacity [to see and position atoms] has allowed us to fabricate rudimentary structures of our own design, atom by atom. . . . The possibilities for perhaps the ultimate in device miniaturization are evident."[21] But the future possibilities of nanotech devices are almost modest in comparison to what the successful imaging and positioning of single atoms have also suggested, namely, that this "atom by atom" fabrication of "novel structures that would otherwise be unobtainable" could theoretically, albeit infinitely slowly and painfully, empower us to "reshape the world atom by atom."[22] The unobtainable is suddenly obtainable, and a molecular monomania takes hold.

Eigler and his research team later used an STM to fabricate what they termed a "quantum corral" (figures 10 and 11).[23] The idea of mastering the fundamental stuff of materiality—subatomic particles as well as larger molecules—originated, we are told, from the ability of the STM to create images. Referring to themselves as "artists" to emphasize the extent to which the practice of STM visualization is a way of seeing that is simultaneously a way of creating, making, engineering, the researchers write:

> The discovery of the STM's ability to image variations in the density distribution of surface state electrons created in the artists a compulsion to have complete control of not only the atomic landscape, but the electronic landscape also. Here they have positioned 48 iron atoms into a circular ring in order to "corral" some surface state electrons and force them into "quantum" states of the circular structure. The ripples in the ring of atoms are the density distribution of a particular set of quantum states of the corral. The artists were delighted to discover that they could predict what goes on in the corral by solving the classic eigenvalue problem in quantum mechanics—a particle in a hard-wall box.[24]

It is the "ability to image" at the subatomic level that in itself "created a compulsion to have complete control." Significantly describing the nanoscale world as a "landscape," the scientists fashion their work as a *mapping practice*, an effort to contain novel territory within a representational topog-

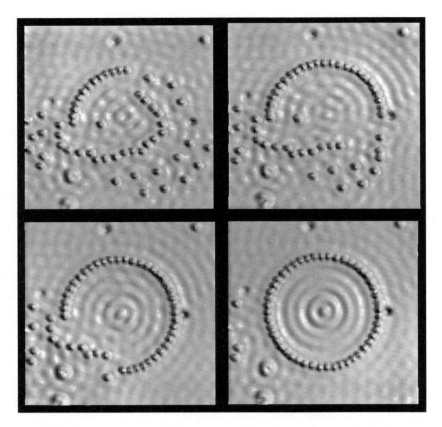

10. Making the quantum corral. This sequence of STM images documents the laborious construction of the circular corral, as reported in Crommie, Lutz, and Eigler, "Confinement of Electrons to Quantum Corrals on a Metal Surface." Courtesy of IBM Research, Almaden Research Center.

raphy that is pictorial, rhetorical, and numerical all at the same time—a "data map," a visual rendering, and a descriptive survey of the landscape that transforms its various *physical properties* into *property as such*.[25] The success in representing and thereby predicting the quantum states of the corral, facilitated by imaging this map in conjunction with the theoretical paradigm of quantum mechanics—in its most "classic" and problem-solving mode as textbook science, reconfirming the very theory presupposed in the architecture of the instrument—convinces the scientists that their compulsion for dominance has been achieved.

Moreover, the landscape is transformed into a physical formation that, in its being *imaged*, comes to be *imagined* through a motif of the American

11. Quantum corral. This color-enhanced STM micrograph of the quantum corral, giving an "atom's-eye" view of the structure, appeared on the cover of the issue of *Science* wherein Eigler's team published their pathbreaking work. Courtesy of IBM Research, Almaden Research Center.

historical experience of westward expansion: the icon of "complete control of the landscape" here is a "corral." The corral as an enclosure of livestock or a circling of the wagons against the wilderness metonymizes the frontier logic of domestication through containment. Indeed, a vocabulary of western exploration and "Manifest Destiny" plays a powerful epistemic role in nanoscience research.[26] Regular descriptions of the molecular landscape as a "new world" and science as encroaching into this promising "frontier" reveal how nano is envisaged not only as a means for containing the molecular world but also as yet one more step in a manifest destiny of human being toward control of the world as a whole through a widening of the visual field.

For example, the nanotechnologist Gregory Timp writes that nanoscientists are "explorers of a new frontier; a frontier that exists on the head of a pin, as incredible as that may seem. These explorers are . . . motivated by curiosity, and the promise of intellectual and monetary rewards, to use science and their powers of observation to map the *terra incognita* of a microscopic world."[27] Seeing this molecular world enables—indeed, "compels"—the scientists to restructure it, to corral it. The exposure of its landscape to the eye and the "powers of observation" makes what was once unseen and wild—beyond the horizon of our human domain—now thoroughly explored and conquered. As Gerard Milburn richly describes it:

> Like the horizon, apparent limits to the progress of modern technology fade before us as we move. The journey down to smaller and smaller scales has, until now, revealed a landscape well described by familiar concepts. But just beyond the horizon lies a new world, the coastline of which has been sighted. Within a decade, high technology will be exploring the unfamiliar world of the quantum.
>
> The world revealed to us in quantum theory is stranger than anything imagined by the wildest mystic, but we understand enough to see how it might be controlled.[28]

With the outlines of this new world "sighted"—visualized and targeted—and with our quantum theory that helps us "understand enough to see," what was once irreducibly strange or unimaginable, undreamed by even the "wildest" mystic, is no longer so. As we journey downward through successive scales, the landscape becomes increasingly "revealed" to our eyes; we push past the horizon, taking the forces of familiarization with us to the very coastline of the formerly veiled and alien world, and in our incremental approach, opaque and apparent "limits" themselves "fade" under scrutiny. Nanoscience directs its gaze against these vanishing barriers and "focuses on perhaps the final engineering scales people have yet to master."[29] With this terminal focus, nanovision exposes and encloses the quantum environment within an engineered "world picture" that makes the *Unheimliche* impossible; everything becomes *heimlich*—everything becomes shockingly *humanized*—when the darkest corners of the world are brought to light and set before our eyes. As Martin Heidegger has written, "The fundamental event of the modern age is the conquest of the world as picture. The word 'picture' [*Bild*] now means the structured image [*Gebild*] that is the creature of man's producing which represents and sets before. In such producing, man contends for the position in which he can be that particular being who gives the measure and draws up the guidelines for everything that is."[30] The ocularcentrist, colonialist project of mapping territory makes the unknown appear known, makes the wilderness appear tamed, makes the distant appear close, reduced to a visual image.

The expansion of the world picture, the gigantic inflation of the object sphere characteristic of modern science, as Heidegger suggests, "evidences itself simultaneously in the tendency toward the increasingly small. We have only to think of numbers in atomic physics. The gigantic presses forward in a form that actually makes it seem to disappear—in the annihilation of great distances by the airplane, in the setting before us of foreign and remote worlds in their everydayness."[31] The containment of the colossal by the velocity of the "increasingly small," the virtual disappearance of distance through miniaturization and speed, the increasingly economic storage of the world into ever more constricted packets of representational space, would therefore seem to culminate with the advent of nanoscale imaging technologies, which focus on the so-called "final scales" of materiality, operating at what nanoscientists regularly refer to as "the limits of fabrication."[32]

In this stock expression from nanodiscourse, "fabrication" must be understood in both its senses as making and making up, building and representing, *techne* and *poiesis*, for in the case of probe microscopy, representing and building are the same: fabrication in the general sense.[33] Gerd Binnig and Heinrich Rohrer—who received the Nobel Prize in Physics in 1986 for their development of the scanning tunneling microscope in 1981 (figure 6)—have written that "scanning tunneling microscopy . . . was the first of a growing family of local probes for imaging and measuring, which can serve at the same time as tools."[34] Functioning as a nanoscale technology of general fabrication—imaging, measuring, and building—the STM comes to fulfill a prophecy: "Besides imaging, [the STM] opens, quite generally, new possibilities for experimenting . . . and [the capability] to modify individual molecules, in short, to use the STM as a Feynman Machine."[35] During the production of an atomic image, the probe microscope can also grasp the atom under scrutiny and manipulate it, and in the process the STM becomes a Feynman machine—the hypothetical device for engineering with individual atoms foretold by Richard Feynman in his 1959 speech "There's Plenty of Room at the Bottom" (owing a significant debt, as we have seen, to Robert Heinlein's science fiction invention of the "waldo"). From relatively early in the history of probe microscopy—for example, during Binnig and Rohrer's acceptance speech for their Nobel Prize on December 8, 1986, or in the writings of the nanotechnologists Stuart Hameroff and Conrad Schneiker[36]— the STM was put into the symbolic role of fulfillment, the coming-into-being of an already inevitable "small-world technology" that was prefigured long ago. As Eigler authoritatively put it in 2001:

On December 29, 1959, in a now-famous and extraordinarily prescient talk titled "There's Plenty of Room at the Bottom" given at the annual meeting of the American Physical Society, Richard Feynman spoke of the possibilities afforded by miniaturization. In that talk he discussed a "great future" in which "we can arrange the atoms the way we want." Feynman's "great future" arrived in 1989 with the discovery of ways to manipulate atoms with the Scanning Tunneling Microscope.[37]

Eigler's own 1989 scriptural IBM experiment, his writing with the STM stylus, would therefore be the New Testament immanence, the material incarnation, of what had been prefigured by the Old Testament of Feynman.[38] An avatar despite himself, Eigler both inherits Feynman's legacy and becomes possessed of Feynman's visionary spirit—or perhaps was always already possessed, unwittingly—an anxiety of influence produced only in retrospect, a revisionist genealogy where the nanotechnological past inherits from the future as much as the converse: "I didn't know about [Feynman's] talk until after I got into the atom-moving business. . . . [But] I felt the ghost of Feynman behind me while I was reading [a transcript of the talk], saying, 'Look, I thought of these things 30 years ago.'"[39]

With a similar sense of inevitability, Eric Drexler has repeatedly affirmed these IBM laboratory experiments with the STM not only as fulfillments of the "top-down," waldo-style method of molecular manufacturing prophesied by Feynman—proof that Feynman's "great future" did indeed arrive in 1989—but also as prefigurations of a more radical "bottom-up" kind of molecular nanotechnology of the general assembler to come "within a few years."[40] The National Nanotechnology Initiative has recounted Eigler's experiments as exactly "the kind of submicroscopic manipulation that Feynman was talking about," claiming that, accordingly, "laboratory feats like these suggest such visions [of the future] are more than mere fantasy."[41] The historical absorption of scanning probe microscopes (along with the microscopists who operate them) into the field of nanotechnology has continually involved this kind of retrospective reframing of SPM research as an integral, even destined, instance of the evolution of nanotechnology and its innumerable anticipated successes.[42] And while many probe microscopists have resisted being lumped into the kind of visions associated with Feynman and Drexler, the appeal of such a move is nevertheless clear: if the STM really is the anticipated "small-world technology" made flesh, then the nanofuture really is already present, and the "reshaping of everyday life" well under way.

Even at that early moment in the history of nanotechnology, Feynman suggested that seeing atoms would enable us to remake the world. "What good would it be to see individual atoms distinctly?" he asks. Well, then "it would be, in principle, possible (I think) for a physicist to synthesize any chemical substance. . . . Give the orders and the physicist synthesizes it. How? Put the atoms down . . . and so you make the substance. The problems of chemistry and biology can be greatly helped if our ability to see what we are doing, and to do things on an atomic level, is ultimately developed—a development which I think cannot be avoided."[43] In the world of this unavoidable future, seeing and doing, visualizing and making, representing and intervening—all are one when rendering the molecular landscape. And, as Feynman promised, by the time we can render this landscape both visually and physically, that is, by the time "we get to the very, very small world . . . [we] will have figured out how to synthesize absolutely anything."[44]

Nanotechnology, working in the "very, very small world" at the "ultimate limits of fabrication," therefore in many ways marks the terminus of the age of the world picture by reducing the worldness of the world to a molecular image, by seeing the wholeness of "our world" and "our future" in its pictures of the "bizarre world of the quantum." Small wonder, then, that in 1990 at the IBM Almaden Research Center, scientists under John Mamin generated an STM image of the Western Hemisphere with deposits of gold atoms (figure 12).[45] The symbolic value of this world-capturing experiment is more lucrative than even the gold filigree might suggest, for it implies that if the globe itself is what is ultimately at stake in the technoscapes and dreamscapes of nanotechnology, certain visionary countries and corporations may already have claimed territory in the small world to come—indeed, they may already have founded the global empires of the nanofuture: "The quest to master the nanoscale is becoming a global competition. . . . Whoever becomes most knowledgeable and skilled on these nanoscopic scales probably will find themselves well positioned in the ever more technologically-based and globalized economy of the 21st century."[46] "Shaping the world atom by atom" is no mere hyperbole, but rather the tropic protocol that scripts the encroachment of the molecular map over the economies and ecologies of the globe: as in the Borges story, this is a picture that becomes the world itself, drawing everything inside.[47]

Nothing escapes. Indeed, nanoscience would seem to complete the world picture by bringing to light those remote spaces that would otherwise fundamentally evade visibility by existing below the Rayleigh limit of optical resolution (roughly half the wavelength of illumination).[48] The nanotech-

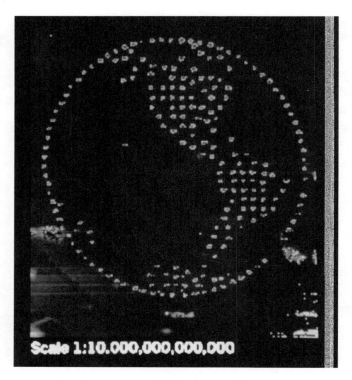

Scale 1:10,000,000,000,000

12. "Gold-Dot Map of Western Hemisphere." Originally published in Mamin et al., "Gold Deposition from a Scanning Tunneling Microscope Tip," this image was later projected behind President Clinton during his kickoff speech for the National Nanotechnology Initiative in 2000. Image courtesy of John Mamin. Reproduced by permission of IBM Research, Almaden Research Center.

nologist's ability to picture the nanoscale world, this ability to capture and map the molecular landscape which by nature cannot be resolved with normal light radiation, this fait accompli of trapping atomic matter within the range of the visible field that humans have evolved to claim as their own, means that nothing can remain hidden from the panoptic scrutiny of nanoscience. As Gerard Milburn has written, "In the basement of my department at the University of Queensland are held captive a small collection of rubidium atoms, trapped in an ultra-cold prison of light and forced to give up the secrets of the quantum world."[49] The quantum world is captured at the moment when our visual range of perception, the wavelengths of radiation to which our eyes are adapted, encircles it and demarcates it as a world, a territory whose secrets are no longer secret once subjected to the scopic regime of modern science. Forced to reveal itself, represent itself, amend its "strangeness" to "our world," the atomic frontier becomes domestic space through this collapse of distance that makes otherness and alterity into the "everydayness" of the visible. The world as such is imprisoned through the deployment of what Paul Virilio has called "subliminal light of incomparable transparency, where technology finally exposes the whole world."[50]

The light that filters through the disciplined eye and its "powers of observation" conquers by simultaneously exposing and establishing a distance. The nanotechnological world picture records the capture of the molecular real by the "prison of light," which surrounds and displaces, puts away in locking away the ineffable or unknowable, the unimaginable. It exposes only to dispose. As Luce Irigaray has said, "More than any other sense, the eye objectifies and it masters. It sets at a distance, and maintains a distance."[51] Or as Donna Haraway has said of this "conquering gaze from nowhere": "The eyes have been used to signify a perverse capacity—honed to perfection in the history of science tied to militarism, capitalism, colonialism, and male supremacy—to distance the knowing subject from everybody and everything in the interests of unfettered power."[52] The STM enacts this double movement of capture and displacement in its production of atomic images and manipulation of atomic structures, for it fabricates the nanoworld and sets the molecular landscape at the distance of the visually rendered data map that, as representation of surface topography, places the very small world away from the viewing subject as an object of mastery. The STM, an instrument of nanovision, is therefore at one level a key apparatus of technoscientific exposure and the scopic regime of the picture.

And yet, at the same time, the STM does something radically different. In shrinking the world to the limits of fabrication—and even, as we will see, beyond the very imagination of limits—its digital images no longer (or at least no longer with ease) set the world at a distance. In fact, the STM and the metaphors used to describe it—or rather, the operations of the tropic protocol that take advantage of instrumental malleability and thereby influence its effects—rely entirely on a *collapse of distance*, an *absolute closeness* of the human with the increasingly small world. This shrinkage of the small world into the human subject enables at a deeply embodied, affective level an alternative trajectory of nanovision away from mastery and toward something else like responsibility and eroticism, or an ethical pleasure in and for the other—the very small other of the nanoworld. While at one remove nanovision performs the culmination of the world picture, it also overcomes the world picture in exceeding the limits of vision that facilitate mastery. For with the demolition of limits, the relation of mastering subject and mastered object can no longer be sustained as a clear break across a chasm of distance and difference.

Nanoscience sees itself as already in control of matter because nano-vision—approaching an infinite smallness of vision, or singularity—collapses the distance between the human world and the molecular world. Yet in this collapse of distance, control itself begins to destabilize. Nanovision makes the world small and engineers the future by taking the scanned probe micrograph as a map of the molecular landscape already colonized; but in occupying this colonized space, the subject of nanovision—the technoscientific viewer—no longer remains the same. The relation of power between the viewing subject and the "prisoners of light" fails dramatically, performatively, and opens to revision at the moment the human dissolves into the molecular frontier.

BEYOND THE MOLECULAR FRONTIER (TO BOLDLY GO . . .)

The cover image of the U.S. National Science and Technology Council's document *Nanotechnology: Shaping the World Atom by Atom* (1999) showcases molecular visualization as a territorial practice extending the frontier of both knowledge and property (figure 13). In the text, Richard Smalley tells us that nanotechnology is "the builder's final frontier," and the iconic coupling of an STM micrograph of a silicon crystal (the "atomic landscape") with the expanses of outer space suggests, as the document puts it, the "vastness of nanoscience's potential," its transformative extension across horizons of every known scale, from the microcosmic to the galactic.[53] (And it's heading straight for Earth!) Our mission, according to Smalley, is therefore clear: "to move boldly into this new field."[54] Similarly, we might compare an analogous image of the molecular frontier, fabricated by the nanoscientist Wolfgang Heckl: a scanned probe micrograph of the nucleobase guanine superimposed against the cosmos, spreading toward the stars (figure 14). This literalization of inner space—the nucleic center of biological bodies—coming to colonize outer space, our visual knowledge of genetic material and our innermost selves (or cells) sprawling to the farthest reaches of the universe, makes palpable the introspective gaze of nanoscience that, in seeing small, resolutely pushes out the external limits of "our world."[55]

So Frederick Jackson Turner seems to have been quite premature in proclaiming the closure of the frontier at the end of the nineteenth century. (Even *Star Trek*'s discovery in the 1960s of outer space as "the final frontier" underestimates the propensity of human beings to generate frontiers or

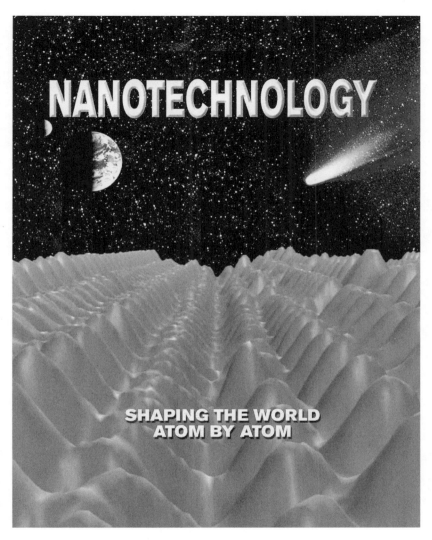

13. "Shaping the World Atom by Atom." Cover image to the National Science and Technology Council brochure *Nanotechnology: Shaping the World Atom by Atom.*

14. "Guanine," by Wolfgang Heckl. Rendered in rich shades of purple, a scanned probe measurement of guanine maps the final frontier, the crest of the sublime. Courtesy of Wolfgang Heckl.

horizons to orient their ever-changing ambitions.) The frontier remains a conceptual driving force of imperial expansion and terrestrial development, even in the era of globalization.[56] The frontier marks the edge of the map and the limit of knowledge as such, and in generating a transcendental beyond as a void yet to be visualized, it produces a scopic drive to occupy and survey that empty space. The frontier establishes a gap in the symbolic field, beyond which lies the real world yet incomprehensible—the real of radical exteriority that Jacques Lacan famously defines as ineffable: "The real, or what is perceived as such, is what resists symbolization absolutely."[57] This symbolic gap itself becomes available to the imaginary only as a site of unsatisfied longing or limitation; the frontier thus has the status of the Lacanian *objet petit a*, the object-cause of desire. The hidden terrain of the real, cut from symbolization and unglimpsable to the imaginary—the "impossible real" whose exteriority Slavoj Žižek therefore casts as "sublime"—remains an inhuman otherness projected outward from the frontier, always beyond reach.[58] Preventing closure of the desiring circuit, the sublime beyond of the intelligible and the visualizable becomes a subjective lack that inaugurates the scopic drive. The existence of the frontier, the discovery of a limit to visualization, therefore produces the world beyond as that which remains uncolonized, unoccupied, unseen and therefore a wound, an opening that Don Eigler, in describing the *molecular* landscape, has called "terra incognita":

> The technological driving force to build things using atomic construction techniques comes from the anticipated economic and technological benefits that it might offer. . . . An equally important and more achievable motivation to build things with atoms is that it allows us to do science in an arena we have never before entered—and that, at least for the scientist, is reason enough. The ability to fabricate on the atomic scale allows us to conduct experiments in a way which here-to-fore was impossible. This ability can best be viewed as a tool which the scientist can use to explore intellectual *terra incognita*. It is the excitement that comes from the exploration that ultimately is the real driving force for building things with atoms.[59]

The "technological driving force" of nanotechnology thus pales in comparison to the "real driving force" of the molecular world itself, the "impossible" as such and the "excitement" it produces, the frisson of "entering" and "exploring," a scientific desire to probe the virgin land, this "arena we have never before entered," with the scientist's "tool" of fabrication. The unknown landscape beyond possibility is itself the object-cause of desire ("that, at least for the scientist, is reason enough"), and the excitement to probe this small world is about the promise of seeing what cannot be seen, inasmuch as fabrication at the atomic scale focuses a certain way of looking at things: "This ability can best be viewed as a tool which the scientist can use to explore intellectual *terra incognita*." Viewing this tool as an instrumental survey of feminized terra incognita therefore sees the advent of nanotechnology as *overcoming the impossible real* . . . in other words, effing the ineffable.

For even in finding the molecular frontier as the limit of visual perception, nanovision tunnels through it. It does so not by extending the optical field (which would only relocate the frontier elsewhere, the "final frontier" simply pushed outward once again), but by employing a new form of nonoptical vision, a new way of seeing that is able to witness and to touch the "here-to-fore impossible." Nanovision describes this way of seeing instantiated by nonoptical technologies of molecular visualization like the STM that, even in discovering the frontier and the limit, open them toward the otherness of the beyond, removing the barrier of perception and entering, instead, what reviewers of the U.S. National Nanotechnology Initiative call the world of "small wonders, endless frontiers," frontiers without limit or boundaries, making nothing but frontiers . . . frontiers that have now come to envelop in their limitless expansion both the human subject and the wonders outside its reach.[60]

"Small wonders, endless frontiers": the conjunction of these two tropes as the essence of nanotechnology suggests significant epistemic transformations. In 1945, Vannevar Bush could speak of science as "the endless frontier," advancing a view of scientific exploration as an eternal project of progressive discovery and opportunity, a vast territory of ceaseless mapping whose stepwise approach to the real would always be asymptotic.[61] But nanotech's appropriation and multiplication of Bush's metaphor embodies a proleptic view of technoscience's *accession* to the real: the sense that the impossible has been made possible, that the future has been enveloped by the present, that space and time have been captured and controlled by the presencing and immediacy of nanovision. That we are already beyond the end. For if nanotechnology purveys and surveys "endless frontiers," a transformation of an infinite singularity into an infinite multiplicity, then the concept of limit has cavalierly been surpassed, for every "apparent limit" would already have vanished simply in this discovery of an eternal succession of ever more frontiers. Moreover, because each successive frontier would itself be endless (infinite in number and infinite in space), the impossible beyond of each endless frontier would itself already be contained within every preceding frontier: the infinite would already be found inside the finite, inside the space of exploration that expands to contain the paradoxical space outside each succeeding one. These endless frontiers extending across each other, a succession of limits enveloping each other in their infinitude, would reveal that there is no sublime beyond, no impossible space, nothing at all outside the interzone of nanotechnology's frontiers because that very impossibility—the real itself—would already have been found inside the arena, the beyond of an other endless frontier necessarily already here within this one.

In the fantasy of traveling through scales that stages the exploration of these endless frontiers—for example, in Gerard Milburn's aforementioned travelogue of a "journey down to smaller and smaller scales"—the experimental systems and techniques of observation available to the emerging nanosciences are assumed to have already extended outside their current dimensional limits. Analogous to the future anterior tense adopted by nanowriting (its science-fictional prolepsis), this perspectival position of the nanovisual gaze is not so much a detached omniscience as a view from the inside that already contains its outside. In Vannevar Bush's idiom of science as a unified "endless frontier," knowledge would be a ceaseless asymptotic approach to the "limits of fabrication"; but in the postmodern nanovisual world of "endless frontiers," each scale already contains the beyond of the next, and the limits of fabrication have already been touched and exceeded. Which is why Baudrillard has said of our technoculture: "It would seem then that there will be no end because we are already in excess of ends: the transfinite. And in an exceeding of finalities: transfinality."[62] Indeed, we might say instead: the *transfrontier*. The outside is already captured by the inside (outer space mapped by inner space, the stars conquered by the image of our nucleic acids). "The 'scientific real,' therefore," Hans-Jörg Rheinberger has written, "is not that ultimate referent to which all knowledge must finally accommodate itself," but rather something that emerges in the overlap of experimental systems (or endless frontiers), which now "are 'more real' than reality": hyperreal networks that produce the real as their own internal limit of fabrication.[63]

The endless frontiers mapped by nanovision thus distill the sublimated real, which condenses—settles and shrinks—into the technoscape. The noumenal now appears in the space of the phenomenal, fabricated as a technological artifact; the "impossible real," even in its radical exteriority, its absolute ineffability, materializes within the nanolab.[64] (Recall that Terry Michalske of the Center for Integrated Nanotechnologies attested to this with paradoxical clarity: "We can really do things we can't imagine right now.") And this is why the many sublime objects encountered in this arena are indeed "small wonders"—small objects that heretofore were wondrous, because unseen and incomprehensible, but whose wonder is now diminished, their sublime features tamed.[65] Nanowriting's narration of the "bizarre world of the quantum"—the weird atomic landscape it regularly describes as a "Wonderland" (with nods to Lewis Carroll) to conjure the cognitive estrangement often experienced in the face of the alien—habitually occurs in tension with a cognitive familiarization, or a resignation that this "here-to-fore impossible" world loses its wonder when it becomes, after all, merely "our world."[66] Even the sublimity of imagined nanofutures or evocative representations of nanoscale objects—the exhilaration of thinking and seeing nanotechnologically—dwindles once the future is rendered present.[67] But this is an expected consequence of the collapse of science with science fiction: the "sense of wonder" which, for classical science fiction, depends on a disconnect between cognitive environment and diegesis, can no longer be sustained when there is no meaningful difference between speculation and mundane reality.[68] Rather, when this happens, a sense of wonder emerges only in equilibrium with a sense of the quotidian. Hence the small-world effect of nanovision, the double movement of the speculative and the specular, fabricates sublime objects and sublime futures only to at once evacuate them of wonder:

> One day in the not-so-distant future, the current interest in nanotechnology will seem quaint. Not because nanotechnology will have failed and vanished, but because it will be so ubiquitous. . . . In engineering terms, there seems so far nothing we can do on the macroscale that we can't do at the nanoscale. . . . That's not to say that nanotechnology is no more than macrotechnology writ small. Some things are very different down at the scale of molecules and cells. Fluids start to look grainy and highly viscous. Surface tension becomes a dominant force. Quantum effects come into play, and everything is buffeted by Brownian motion. But with a bit of ingenuity, all of these things can be used to advantage. . . . Small wonder that companies such as IBM are starting to look seriously at nascent technologies, including molecular self-assembly, carbon nanotubes and other nanowires, and scanning-probe devices. . . . Which is why, for some researchers, nanotechnology (as opposed to exploratory nanoscience) is at present very much an extrapolation of the familiar, an evolutionary rather than revolutionary affair.[69]

Nanotech wonders become "quaint"—small, feminized, and rustic—once science and its corporate developers begin to take this stuff "seriously." That is, once "science fiction" has been purged from "real science." There is no place for wonder in the face of the "serious," no place for the "revolutionary" once the familiar and the present have extended across the "very different" nanoworld as well as the "not-so-distant future." It would seem that as our perceptions of nanotechnology become increasingly "ubiquitous" to the point of invisibility—the "growing (shrinking)" characteristic of the nano world picture—no place for wonder remains at all. ("Small wonder that . . .") The space of rare experience, the space of the sublime, gets swamped by the ubiquitous and the commonplace—anticipated already, "one day," even before nanotechnology has in fact become ubiquitous. Binnig and Rohrer, who carefully negotiated the novelty of their atomic-resolution STM images at the IBM Zurich Research Laboratory (a place that takes new technologies "seriously"), claim that we have already become blasé about nanowonders: "The appeal and the impact of STM lie not only in the observation of surfaces atom by atom, but also in its widespread applicability, its conceptual and instrumental simplicity, and its affordability, all of which have resulted in a relaxed and almost casual perception of atoms and atomic structures."[70]

Certainly, this kind of casual domestication of the sublime has histori-cally accompanied territorial movements at the frontier of visibility. The his-tory of light microscopy, for example, from at least Robert Hooke onward, has long evinced a fascination with the sublime; but as Mary Baine Camp-bell shows, the microscopic sublime deflates at the moment of its produc-tion, digested and processed as any mundane object as soon as it appears.[71] So too throughout the history of the electron microscope, which once re-vealed an "Alice in Electronland" vista of the unseen—an even smaller world than that disclosed by light radiation—the microscopic sublime not only de-flates once discovered, but often vaporizes entirely.[72] The production of the electro-micrographic image frequently involves annihilation of the sample itself, erasing the unseen object in the act of exposing it: "Indeed, electron radiation experienced by a specimen in the microscope beam is said to be about the same as would be received from a ten megaton H-bomb blast 30 yards away. Biological specimens . . . lose around one-third of their dry mass in the first minutes of viewing, as their substance is stripped away under the intense bombardment."[73] The sublimity of the invisible vanishes even as the electron beam blasts the sample to waste, the wondrous aura of the "origi-nal" scrubbed out as an event of the age of electrographic reproduction.[74]

If modern scientific vision has been compelled to image the unseen as a proper policing of wonder—leaving a wake of rationality and reason behind its perpendicular penetration into the invisible, its pornographic drive to open and expose—then the STM would simply mark another progressive step in the project of panoptic expansion.[75] However, image productions of earlier microscopic regimes, even in recording this incremental recession of the surface of visibility, have only succeeded in reinforcing the positivist imaginary of a singular "endless frontier" whose exterior would always re-main technically invisible—the wildlife preserve of the sublime and won-derful—and therefore new sublime space always appears as an effect of in-creasing distance between viewer and object. As Walter Benjamin famously suggested, the aura of the sublime appears as "the unique phenomenon of a distance, however close it may be."[76] The image productions of the STM, on the other hand, instead of extending the distance of vision, record the outside of the specular surface drawn forth into the visible. This is a dra-matic reversal of the direction of visualization: not a penetration into the unseen but a phagocytosis of the outside. Like other visual transcodings of nonoptical data, the STM image captures what is beyond the very possibility of visible surface and thereby snuffs out all traces of an exterior sublimity. (Lev Manovich has even described such visualizations of data sets as the "anti-sublime.")[77] STM images are algorithmic transformations of raw data, manipulated by the computational system and the investigators to isolate or amplify certain characteristics of a physical sample, imposing a reality principle on the data such that the image literally becomes a better rendition of reality than the data itself.[78] Or rather, the STM image marks one site of data-visualization practice where the real is no longer "out there" but "in here."

STM micrographs convey electronic information about the nanoscale environment while simultaneously simulating that information as a place, as a small world that comes into being through interactions with the instrumental hardware and its software processing. STM images therefore surpass representation and find themselves already outside the frontier: speculations beyond the specular, through the looking glass. They take over the place of the real. And if there are correspondences, they are not so much communication as *passages*. As the nanoresearchers Christoph Gerber and Hans Peter Lang have written: "With the emergence of scanning probe microscopy (SPM) and related techniques in the 1980s, the door to the nanoworld was pushed wide open."[79] The STM image *accesses* the nanoworld as such, less a new visual limit than a graphical interface or tunnel. Which is why, where earlier, optically mimetic micrographs could not avoid perceptions of distance between viewer and microcosm—at the least, the scaled-up analog distance of that traversed by a diffracted photon or a transmitted electron—the image operations of the STM have regularly entailed perceptions of "being there," of a radical loss of separation between viewer and nanocosm.

From the earliest encounters with the imaging capabilities of probe microscopy, this collapse of distance has suggested a presencing or occupation of the nanoworld. The occupation seems to occur as an effect of seeing what has not previously been seen, and what indeed is technically impossible to be seen as such, at least with the human visual apparatus that depends on light. But inasmuch as the probe microscope can produce the molecular world as a hyperreal topography, it encounters the impossible in the form of what Binnig and Rohrer call a "real-space image" (as opposed to the fully imaginary and inverted "reciprocal space" generated, for example, by x-ray diffraction techniques).[80] Even in the first moments of success, the duo found themselves occupying that world beyond the frontier as a result of reconstructing and resolving "real space." As Binnig put it: "I could not stop looking at the images. It was like entering a new world. This appeared to me as the unsurpassable highlight of my scientific career, and, therefore in a way, its end."[81]

In its "appearance," its presencing, this encounter with the new world of molecularity is "unsurpassable" as a scientific experience. It becomes the sublime "end" of scientific experience because, rather than now locating some other "final frontier" elsewhere, situating the always-unfulfilled "real driving force" of scientific desire (*objet petit a*) beyond an endless frontier, the STM opens the impossible real *inside* the frontier as a hyperreal space that has now apparently been "entered." It is unsurpassable precisely because its limitations have been exceeded, and the new world in which the subjective imaginary finds itself when gazing at the nanoscale image has expanded to contain the observer. Or more precisely, the observer *experiences* this impossible seeing across the limits of visibility as entering sublime space and actually touching the real itself. As the nanoscientist James Gimzewski and the nanomedia artist Victoria Vesna have written:

> The new tactile techniques opened up a radically new approach to microscopy enabling real local properties to be imaged and mapped. For instance, ultra high-resolution images of local magnetism like bits of north and south directed domains could be obtained with magnetic tips. If friction was an issue, images of local friction as it scanned the surface could be mapped. This opened up a new world, a world never really seen before on those terms—the nanoworld.[82]

This "new world" is opened by the tactile practices of probe microscopy that generate the image and the map of "real local properties," a survey, as it were, of real estate, local properties inventoried by the property inspector, domestic "domains" with their many attractive or "magnetic" qualities now put up for sale. In viewing these nanoworld pictures, these images of formerly uninhabited local properties, one comes to live there. As the chemist John I. Brauman puts it, "There is, indeed, room at the bottom, and we are beginning to move in."[83]

In Gerard Milburn's account of the "nanoManipulator" developed at the University of North Carolina, Chapel Hill—a device that joins a scanning probe microscope to an immersive virtual reality environment, including a haptic interface in the form of a force-feedback toggle (figure 15)—the digital transformation of data between tunneling current and hand movements enables users to occupy nanoworld real estate:

15. nanoManipulator, University of North Carolina, Chapel Hill. This "workbench" model supplements the original vr Nanomanipulator project directed by Warren Robinett (University of North Carolina, Chapel Hill) and R. Stanley Williams (ucla). Photograph by Larry Ketchum. Courtesy of the Department of Computer Science, University of North Carolina, Chapel Hill.

Our understanding of the everyday world is built up from manifold experiences mediated by our senses. It is often said that quantum mechanics looks so difficult because we cannot have any direct sensual contact with the atomic realm. The nanomanipulator indicates that this may not always be so. Perhaps we can see and touch the quantum world. Like a newborn baby, our interior image of this world would be brought up through action and play. Sitting by a sea of electrons, we could pile atoms on top of atoms in atom-castles. We could cast atoms in the electron sea and watch the ripples move across the surface. We could almost live in the strange world of the quantum.[84]

The physical encounter, the sensual contact, with atomic-scale objects produces the experience of *worlding*, of other worlds built up from exploration, from seeing and touching. The sensorimotor foundations of the worlding event themselves seem to literally embody the tropes of science fiction, rendering them as live action: "Recently, an episode of *Star Trek: Voyager* showed crew members performing genetic engineering simply by touching pictures of genes on the screen of a computer. Yes, it's only television, and science fiction at that, but life is beginning to imitate art at the University of North Carolina at Chapel Hill."[85] The nanoManipulator, nothing less than real science, remains laden with the semiotic residue of the novum: "Researchers can travel over genes, tickle viruses, push bacteria around, and tap on molecules. . . . It seems like something out of Robert Heinlein's 1942 [*sic*] book, *Waldo & Magic, Inc.*, in which the hero of the story learns to manipulate tiny forces with his remote control machines."[86] As the computer scientist Warren Robinett, who created the nanoManipulator together with the chemist R. Stanley Williams, explains: "It is like the [Disney] movie *Honey, I Shrunk the Kids*, except you don't really get smaller. . . . We reconstruct your perception so that you see and feel exactly what you would if you were the size of a virus."[87] The science fiction dimensions of nanotechnology wash together with Disney cinema in the tropical current of the laboratory, shaping the user's engagement with the probe microscope. Operating according to this tropic protocol, the instrument reconstructs perceptual experience, shrinking the sensorium into the nanoworld even while the body remains in our world (that is, you get smaller except you don't really get smaller).

This shrinking of sight and touch that literally opens up another world beyond visibility, inside the self as what Gerard Milburn calls "interior image," is the product of a technologically mediated vision that is no longer optical or ocular. Binnig and Rohrer write: "The STM is a mechanically positioned, electrically sensitive kind of nanofinger for sensing, addressing, and handling individually selected atoms, molecules, and other tiny objects and for modifying condensed matter on an atomic scale. . . . And like with finger tips, it is the 'touch' that makes the difference."[88] Seeing the molecular is enabled by touch, by haptic encounter. It is a way of seeing space beyond the closure of an ocularcentric regime that Gilles Deleuze (following Aloïs Riegl) has called "haptic vision." As Deleuze writes, haptic vision takes place "when sight discovers in itself a specific function of touch that is uniquely its own, distinct from its optical function." It is a "joining together of the two senses of touch and sight, like the soil and the horizon."[89] And space therefore contracts, because if haptic vision joins touch to sight as the soil to the horizon, then long-range perceptions are impossible, for in no sense then can there be grounds for any real outside the immediate, and the spatializing effects of haptic vision constitute an absolute closeness. Deleuze and Guattari write, "Where there is close vision, space is not visual, or rather the eye itself has a haptic, nonoptical function: no line separates earth from sky, which are of the same substance; there is neither horizon nor background nor perspective nor limit nor outline or form nor center; there is no intermediary distance, or all distance is intermediary."[90] Within haptic space, distance does not separate and appears only as the medium of passage, or interface.

Valerie Hanson provides a striking account of the way in which scanning probe microscopy works through haptic vision. She argues that a synesthesia of sight and tactile-motor sensation is produced for the microscopist—and even for the casual viewer of STM images—as a pattern of conditioned response to images understood to have been made possible by the "touch" of the probe tip to the atomic surface, by the quantum mechanical assumptions built into the instrumentation, which suggest that our seeing of the nanoworld emerges only because we are already touching it, manipulating it, and interacting with it.[91] This trope of "touch" is deployed in many nano-rhetorical accounts of probes as fingers, for example, when Eigler writes that the STM "forms an image in a way which is similar to the way a blind person can form a mental image of an object by feeling the object,"[92] or when Binnig and Rohrer depict the local probe as the "gentle touch of a nanofinger . . . [wherein] the two atoms that are closest to each other are able to 'feel' each other"[93] (figure 16). In these examples, "touch" constitutes the tropic protocol for the viewer of nano-images and for the microscope user in generating maps of local properties or in generating data of various surface properties that can be rendered visual or otherwise but, in their many digital transformations, their many graphical representations, are all still conceived as products of the "touch that makes the difference."

The visualization of the nanoworld therefore aboriginally occurs as synesthesia, as haptic vision—as fabrication in the general sense—where the image cannot be separated from conceptions and sensations of the fingering operations of the STM. The nano world picture demands a difference in the way we experience our sense ratios, and in what kinds of information our various senses are able to communicate. The touch of the local probe can convey various pieces of information about the molecular surface—texture, thermal conductivity, magnetic properties, energetic and spatial arrangements of electrons, strength of interaction, and so forth—all of which can be displayed visually as an image, which itself can be transcoded instead as an audio signal, a graph, a numerical matrix, a 3-D object, or whatever: as Binnig and Rohrer put it, "A colorful touch, indeed, with a rainbow of possibilities."[94] Certainly, nanotechnologists have become highly sensitive to the alterations of perception occurring at the interface of the human probe microscopist and the instrument as it touches the real. Don Eigler and Joseph Stroscio have written of this sensory tweak: "In a sense, we may use the STM

The Concept

The Realization

16. The gentle touch of a nanofinger. The STM becomes a digital extension, putting us "in touch with atoms." Reprinted with permission from Binnig and Rohrer, "In Touch with Atoms," *Reviews of Modern Physics* 71 (1999): S324, fig. 1, © 1999 American Physical Society.

to extend our touch to a realm where our hands are simply too big."[95] Others express delight in the sensual or erotic intimacies that probe microscopy enables, for now "we can 'feel' atoms like a finger feels Braille dots. Being this intimate with matter is unprecedented."[96] Or as Gimzewski and Vesna put it, "This new science is ultimately about a shift in our perception of reality from a purely visual culture to one based on sensing and connectivity."[97]

The probe microscope enables an extension of perception beyond the body in a way that is nevertheless experienced in a deeply embodied way. Nanovision occurs at the cyborg interface of the human and the STM as something like a transhuman vision, a hybrid of the machinic and the human nervous system.[98] The touch of the STM and the touch of the human operator work together to produce the nanoworld as haptic space, for in the process the operator actually senses within the meat of his or her body an inhabitation of the nanoworld as an effect of probe-data image production. As Gimzewski and Vesna write:

> The Scanning Tunneling Microscope represents a paradigm shift from seeing, in the sense of viewing, to tactile sensing—recording shape by feeling, much like a blind man reading Braille. The operation of a STM is based on a quantum electron tunneling current, felt by a sharp tip in proximity to a surface at a distance of approximately one nanometre. The tip is mounted on a three dimensional actuator like a finger. . . . This sensing is recorded as the tip is mechanically rastered across the surface producing contours of constant sensing (in the case of STM this requires maintaining a constant tunneling current). The resulting information acquired is then displayed as an image of the surface topography. . . . Through images constructed from feeling atoms with an STM, an unconscious connection to the atomic world quickly becomes automatic to researchers who spend long periods of time in front of their STMs. This inescapable reaction is much like driving a car—hand, foot, eye, and machine coordination becomes automated. Similarly, the tactile sensing instrument soon became a tool to manipulate the atomic world by purposefully moving around atoms and molecules and recording the effect, which itself enabled exploration of interesting new physical and chemical processes on a molecule-by-molecule basis.[99]

A remarkable description in many ways. It at once expresses the way in which media technologies come to act as what McLuhan called "the extensions of man," expanding the range of the body and the sensorium through medial prosthesis, as well as the blurring of boundaries at the splice site of human and machine, for example, the way in which the vehicular operator comes to experience the parts of the car as reticulations of the body, a biomechanical assemblage wherein "hand, foot, eye, and machine coordination becomes automated."[100] The human is made cyborg—or rather, its innate cyborg potential activates—when plugged into the machinic extension.[101] Such a fate has become increasingly ubiquitous in the media landscape described by McLuhan, or in the technoculture fantasized by J. G. Ballard's *Crash* (1973), where the human body and imaginary have become so interpenetrant with the technoscape that there is no longer an absolute difference between the car and its driver, or a car crash and an orgasm.

While suggesting that this cyborg extension of the human microscopist occurs as a literal extension of the sensorimotor structure right up to the limits of fabrication (echoing sentiments throughout nanodiscourse that "the STM has become our hands and eyes to explore the quantum world"),[102] Gimzewski and Vesna also get at the way in which the new media technology of the probe microscope collapses distance between the human subject and molecular space: "Through images constructed from feeling atoms with an STM, an unconscious connection to the atomic world quickly becomes automatic to researchers who spend long periods of time in front of their STMs." We begin to see that the instrument, when grafted to the human body, changes human perceptual ratios so utterly that the entire psychic structure of the subject shifts. Now instead of finding the real as the impossible beyond of the symbolic and imaginary fields, the subject finds the space of otherness inhabiting the self.

Friedrich Kittler has argued that, in the media ecology of modernity, the human psychic apparatus has been divided and externalized among the materialities of communication such that the orders of the symbolic, the imaginary, and the real are incarnated as technologies of writing, optics, and acoustics—that is, by the technical registration and storage of the machinic symbolic, the visual imaginary, and the unmediated flux of the real. Hence in the subjectivization processes of "so-called Man" in the modern world:

> The typewriter provides writing as a selection from the finite and arranged stock of its keyboard. . . . Thus, the symbolic has the status of block letters. Film was the first to store those mobile doubles that humans, unlike other primates, were able to (mis)perceive as their own body. Thus, the imaginary has the status of cinema. And only the phonograph can record all the noise produced by the larynx prior to any semiotic order and linguistic meaning. . . . Thus the real . . . has the status of phonography.
>
> Once the technological differentiation of optics, acoustics, and writing exploded Gutenberg's writing monopoly around 1880, the fabrication of so-called Man became possible. His essence escapes into apparatuses. . . . So-called Man is split up into physiology and information technology.[103]

Effected by the historical differentiation of media, human(ist) subject formation would therefore be localized to a particular media ecology, a particular era, and prone to revision according to material changes in medialization. And indeed, it would seem that the medial and psychic divisions which have heretofore served to produce what we know as the human subject are all reconfigured in the economic repackaging of the "extensions of man" and perceptual ratios by the STM. If psychoanalytic categories have been heuristically useful for an understanding of "so-called Man" and humanism in the modernity of analog media, their reassembly by nanotechnological instrumentation points beyond the human toward possible constructions of the posthuman in digital technoculture—and likewise the need for an upgraded analytics adequate to the (de)subjectivation processes mediated by nanovision.

The psychic apparatus of the probe microscopist becomes deployed differently: a technical reconfiguration of the human in its cyborg interface with the haptic-vision machine. In viewing the algorithmic images produced by the STM and physically operating the probe tip to scan the field of the molecular object, the microscopist inhabits a new media ecology. "Through images constructed from feeling atoms with an STM [the nanoscale image, misidentified by the human observer as inhabitable space where "we are beginning to move in," constitutes the imaginary] an unconscious connection [the symbolic, the translationary data field, the machinic processes that generate and reconfigure tunneling current into information] to the atomic world [finally, the real, the sublime beyond, the molecular world itself, the electron cloud and the tunneling particles as they interchange with the probe tip] . . . [etc.]." The media ecology of the STM — nanoscale haptic image, machinic symbolic, and the real interchange of electronic particles tunneling across the quantum "vacuum gap"[104] or "forbidden space"[105] between the molecular world and the STM's tip, outside processed data—comes to instantiate, to materialize, the human subject's perceptual ratios differently, such that the encounter with the world is apprehended now not as a division of perception (for example, into visual or auditory ratios), but rather as quantum connectivity, or entanglement, where "real space" is touched across the symbolic translations of data, across what Gimzewski and Vesna call the "unconscious."

The experience of the impact of the real, outside consciousness, "quickly becomes automatic to researchers who spend long periods of time in front of their STMs." It is an affective moment, an unconscious, as-if-unmediated experience of direct "connection" with molecularity. In the same way that new media work to make it seem that no mediation is involved in presenting an immersive immediacy—what Jay David Bolter and Richard Grusin call the new media ideal of "transparent immediacy," or the "transparent interface . . . that erases itself, so that the user is no longer aware of confronting a medium, but instead stands in an immediate relationship to the contents of that medium"[106]—viewing the images produced by the STM while operating the machine makes it seem as if the human is in direct confrontation with the molecular nanoworld. It's not, of course, that the operator is "actually" touching individual atoms (however that would differ from our everyday encounter with gross matter), but rather that the medial translations and remediations of electronic convergence have produced sensations of connectivity at a radically different scale: tunneling all the way down, with no uncrossable or "forbidden" gaps between our world and the world of the quantum. Thus the real local properties of the nanoworld become interdigitated with the exploded fingertips of the posthuman colonist. The creators of the nanoManipulator put it this way: "We are building *visualization systems* that intuitively map the additional senses made available by these [scanning probe] microscopes into the human senses, and *control systems* that project human actions directly into this world. . . . [This puts] scientists virtually *on* a nanometer-scale surface *in* direct control *while* experiments are happening."[107]

The human touches the nanoworld, or at least experiences a sensation of "projecting" the body and its actions into this world, precisely because of the reconfigured media environment of the probe microscope. While so-called old media preserve a fundamental analog distance between the human sensorium and the real, new media technologies of data visualization strive to enable an experience of distanceless encounter. As Mark Hansen has suggested, "If digitalization underwrites a shift in the status of the medium — transforming media from forms of actual inscription of 'reality' into variable interfaces for rendering the raw data of reality — then . . . the very task of *deciding* what medial form a given rendering shall take no longer follows from the inherent differences between media (which have now become mere surface differences). . . . Simply put, as media lose their material specificity, the body takes on a more prominent function as a selective processor of information."[108] Indeed, the body comes to *frame* the rendered data image. Which is why the impossible beyond of the visualized information occurs as a haptic space of *interiority*, the spacing of the subject within itself, an affective response of "bodily spacing or the production of space within the body," that is, the sensation of "worlding."[109]

This internal spacing, articulated by those who work with probe-microscopic media as an "unconscious connection to the atomic world," or as an "internal image of this world brought up through action and play," is the production of another space, an alien space — the brave new nanoworld — as a sensory experience inside the body. A touch of the electronic real has become instrumental data rendered by the wish-fulfilling nanoscale imaginary, and it remains undecidable where to draw a division, where to mark a discontinuity when matter is becoming-data even as data is becoming-flesh. This undecidability figures as a unique feature of the tactility of the STM, which, as an extension of the human body, merges with the molecular world in such a way that the sublime is domesticated as part of the self, and it becomes unclear where the electronic "extensions of man" end and the nanoworld begins.

The operations of the STM are perceived to produce the nanoscale image precisely because of a tropic protocol of "no distance" between the probe and the sample, the electrons from the nanoworld tunneling across the "touching" and "overlapping" valence orbitals of atoms that make up the STM's tip.[110] This loss of distance in a small world so small that it becomes coextensive with "our world" makes the very imagination of separation or contact between viewer and viewed uncertain, indeterminate. As Binnig and Rohrer write: "The concept of contact—electrical or mechanical—blurs at the molecular scale. In the case of electrical contact, no sharp boundaries exist because of the penetration of electronic wave functions into the potential barriers of finite height, giving rise to electron tunneling. . . . On the other hand, interference and quantum effects can lead to discontinuities like quantum conduction."[111] To be "in touch with atoms," as they put it, as an experience of the interface between human body, imaging probe, and molecular nanoworld, makes it impossible to decide between either continuity or discontinuity of subject and object, self and other, for these concepts themselves "blur" at the nanoscale.

Therefore the imaging of nanoscale objects is experienced as a literal touch, and even as inhabitation of a new world, not only because of the power of the nanoworld picture to suggest that this new world is small and therefore graspable, but because the body processes the tropically mediated image in conjunction with the motive actions required to operate the STM, and these processes produce an affective state of being there, of physically merging with the molecular. To say it differently, nanovision relocates the sublime beyond of the frontier into the data frame, that is, the body. Located in the body as affect, as a spacing of the subject within itself, the impossibly vast nanoworld collapses into the human as an experience of proprioceptive control. For though the molecular world may in fact escape and extend far beyond the formal boundaries of the body, it has been impressed as an inner space, an internal vision of inhabitation, an occupation of foreign real estate, as if the world and the human were coextensive. Under the regime of nanovision, the nanoworld is made present by virtue of the way in which the technical apparatus collapses distance into the body. No longer outside, technology brings itself and the rest of the world inside. There is no distance. And therefore it truly is a small world, after all.

SURFACE TENSIONS

Nanovision is a form of embodied perception that abolishes distance in seeing the very question of continuity and discontinuity as undecidable, or blurred. Nano finds no absolute frontier between worlds, no clear separation between observer and observed, no fundamental split or space between microscale and macroscale, no forbidden quantum leap between the molecular and the molar—and equally, as we have seen, no boundary between present and future, or between science and science fiction. In discovering the outside to already inhabit the inside, nanovision leaves *no space beyond* the "endless frontiers of nanotechnology." There is no "impossible real," no sublime object, no "real space," that cannot be known, that cannot be seen, that cannot be touched, that cannot be inhabited, that cannot be represented, that cannot be constructed. Nanovision moves endlessly beyond limitations: according to the physicist and Nobel laureate Horst Störmer, "Nanotechnology has given us the tools. . . . The possibilities to create new things appear limitless."[112] It is not so much that nanotechnology is science at the limits of fabrication, then, but rather that, physical law notwithstanding, *nanotechnology is the science of fabrication without limits.*

What had previously been seen as limit becomes revisioned as surface tension, an interface of worlds held separate only by a terminal instability. And this interface becomes the blur-space where nanovision occurs. For if nanovision makes tangible the impossible real from within one imaginary field, it simultaneously deprivileges the location of perception in that field; in touching between worlds, nanovision brings one into the other, and the viewing subject of nanovision does not see a different world across space but discovers no space to cross. When worlds collide, it is no longer possible for one world to look at the other from a secure distance, from a godlike vantage point (what Haraway has called the scientific "god-trick of seeing everything from nowhere"),[113] because the human observer has now become situated within the other world even as that other world is situated as a space within the human observer. Which therefore opens the possibility that, when we look at the nanoworld, the nanoworld looks back.

This gaze from the nanoworld turned outward to the macroscale—the same perceptual orientation, for example, that structures the cover of *Nanotechnology: Shaping the World Atom by Atom* as an image of human mastery that simultaneously looks out from the viewpoint of a molecular spectator (figure 13)—spans the history as well as the prehistory of nanotechnology, appearing long before the invention of the scanning tunneling microscope. In 1923, for example, Ray Cummings published *The Girl in the Golden Atom* (combining his stories "The Girl in the Golden Atom" [1919] and "The People of the Golden Atom" [1920]). The central scientist-hero—named only as "the Chemist"—is convinced of the "infinite smallness" of matter: "In other words, what I believe is that things can be infinitely small as well as they can be infinitely large. . . . How can you conceive of the edge of space? Something must be beyond—something or nothing, and even that would be more space, wouldn't it?"[114] Scopically driven by the sense that every scale of smallness contains yet another beyond the horizon of visibility, the Chemist plumbs microscopic depths: "I had never gone in for microscopic work much, but now I let it absorb all my attention. I secured larger, more powerful instruments—I spent most of my money, . . . but never could I come to the end of the space into which I was looking. Something was always hidden beyond—something I could almost, but not quite, distinguish" (2). Even inventing an ultra-microscope of subatomic resolution cannot satisfy his desire to see smaller, to peer into the tantalizing blackness of the beyond: "a blackness that seemed not empty, but merely withholding its contents beyond my vision" (4).

The Chemist eventually discovers a method to shrink himself and physically explore the small worlds of matter. He decides to enter the metallic landscape of a gold ring, voyaging downward through scales and eventually entering a single golden atom. As he thus projects his body beyond blindness, suddenly the ineffable outside becomes an interior—a radical shift of perspective: "In our [macroscale] world here the horizon is caused by a curvature of the earth below the straight line of vision. We are on a convex surface. But as I gazed over this [atomic] landscape . . . I saw at once that quite the reverse was true. . . . There was no distant horizon line, only the gradual fading into shadow of the visual landscape. I was standing obviously on a concave surface, on the inside, not the outside of the world" (40–41). In this early instance of nanovision, sublime exteriority moves inside and vice versa, a transfrontier erasure of the horizon line whereby every surface contains every other, and even "our world" becomes just another small world already beyond the limits of fabrication at another scale:

I believe that every particle of matter in our universe contains within it an equally complex and complete a universe, which to its inhabitants seems as large as ours. I think, also that the whole realm of our interplanetary space, our solar system and all the remote stars of the heavens are contained within the atom of some other universe as gigantic to us as we are to the universe in that [gold] ring. (7–8)

Already in the early decades of the twentieth century, then—before Feynman, before Drexler, before Binnig and Rohrer—imaginary small worlds had begun to unfold according to the architectonics of nanovision. Taking place belatedly, the phenomenology of probe microscopes in interaction with human operators would seem *preconditioned* by the same logic structuring the narratives of certain fictions. Nanovision appears therefore not as a consequence of the technological ability to explore and manipulate atoms, but rather as a function of speculating the possibility of atomic encounter in the first place. In other words, the essence of nanotechnology, as Heidegger might say, is by no means anything nanotechnological; instead, the instrumentation of nanoscience, such as the STM, emerges already enframed by the way of seeing described by nanovision and the tropic protocols it articulates.[115]

The same tropic protocols that undergird the small-world effect of STM usage have also shaped the longer literary history of nano. For example, the transverse journey through scales of infinite smallness, entangling our world with the strange world of the quantum, has become a formal convention for many texts informed by nanovision. In the graphic narrative "Lost in the Microcosm" (1950) by Al Feldstein and William M. Gaines, drawn by Harvey Kurtzman and published in the inaugural issue of the comic book *Weird Science*, two scientists create a technology to "compress matter to an infinitely small size . . . an almost incomprehensible size!"[116] One of them, Karl the postdoc, accidentally exposes himself to the system and begins to shrink, quickly disappearing into the palm of his PI's hand. While the PI, Professor Einstadt, thus touches the microscopic, the younger scientist inhabits it: "And so I left my world and my dimension! Now, everything that I had once squinted at through powerful microscopes became huge and threatening! The flesh of the professor's palm was now a giant, wet, oozing labyrinth, full of canals and passageways . . . throbbing with life!" (14). Battling tuberculosis bacilli and white blood corpuscles inside the professor's hand—prefiguring images of medical nanobots patrolling the bloodstream as well as the iconic film version of this conceit, *Fantastic Voyage* (dir. Richard Fleischer, 1966)—

the shrinking scientist soon disappears into the molecular structure of the cells: "I shall be the first man to look upon a *molecule*!" (14). Karl then sees that each molecule contains its own small universe: "Now I was reaching the frontier of the microcosm, as yet undiscovered by Man! . . . The atom . . . the tiny atom . . . is actually an astronomical solar system!" (15). Discovering that one of the subatomic particles, or planets, "looks habitable," Karl establishes a haptic connection with the ineffable: "As I grew smaller . . . the [subatomic] planet grew larger . . . I reached out and gently touched its surface!" He then shrinks into this particulate world, which turns out, in a twist ending, to be "our world," our own planet Earth: "Can it be that this Earth is just a puny world within a world all somewhere within an atom in the flesh of some professor's palm?" (18). Inside the hand of the scientist, the nanoworld — which is also our world — unfolds. Discovering worlds beyond worlds, Karl continues to shrink forever, infinitely deeper into the hand of his professor, which now contains all possible worlds, even as it is itself contained. In a similar vein, Richard Matheson's novel *The Shrinking Man* (1956), ends with the protagonist, Scott Carey, shrinking endlessly beyond the microscopic:

> Why had he never thought of it; of the microscopic and the submicro-scopic worlds? That they existed he had always known. Yet never had he made the obvious connection. He'd always thought in terms of man's own world and man's own limited dimensions. . . . To a man, zero inches meant nothing. . . . But to nature there was no zero. Existence went on in endless cycles. It seemed so simple now. He would never disappear, because there was no point of non-existence in the universe.
>
> It frightened him at first. The idea of going on endlessly through one level of dimension after another was alien. . . . He stood in speechless awe looking at the new world. . . . It was a wonderland. . . . Scott Carey ran into his new world, searching.[117]

Presaging the "small wonders, endless frontiers" of nanodiscourse, these texts body forth the characteristic protocols of nanovision, finding plenty of room at the bottom — and beyond.

Or consider James Blish's novelette "Surface Tension" (1952). Voted by the Science Fiction Writers of America (SFWA) as one of the fifteen most important stories published between 1929 and 1964 (roughly, the golden age of science fiction), "Surface Tension" takes place in a small world where nanoscale phenomena such as molecular adhesion forces and density gradients constitute the principal features of lived reality.[118] Generations ago, human space explorers from Earth crash-landed on an inhospitable planet possessing only a few small freshwater pools. Recognizing that their days on this planet were numbered and without hope of rescue, the explorers decided to biologically engineer their offspring to survive in the alien environment. Shrunk down to microscopic proportions—less than twenty-five microns tall—the offspring were also endowed with webbed extremities, book-lungs and respiration-spiracles like those of arachnids to allow water breathing, a rapid reproduction cycle, and the ability to sporulate in order to survive the freezing winters by hibernating in a hard protective shell. Thus designed to make the best of the new world, the offspring were released into a freshwater pool to prosper and multiply. But now, after many generations of living in their tiny aquatic world, the microscopic humans have forgotten their ancestry and perceive the boundaries of their pool to mark the extent of the universe. According to their cosmology:

> [There were] three surfaces of the universe.
> The first surface was the bottom, where the water ended.
> The second surface was the thermocline, the invisible division between the colder waters of the bottom and the warm, light waters of the sky. . . . A real interface formed between the cold, denser bottom waters and the warm reaches above, and maintained itself almost for the whole of the warm season.
> The third surface was the sky. One could no more pass through that surface than one could penetrate the bottom, nor was there any better reason to try. There the universe ended.[119]

This top surface of the world, the "end of the universe," is a mirror: "the wavering mirror of the sky" (407). And it functions to mark the limits of the imaginary. Some of the microscopic people might fancy the existence of "other universes besides this one, but where these universes might lie, and what their properties might be, it is impossible to imagine" (400–401). Like the mirror stage in Lacan's account of subject formation, this "specular image seems to be the threshold of the visible world," reflecting back an image of the universe with the subject at the center, the center of all phenomenological positivity: "The function of the mirror-stage . . . is to establish a relationship between an organism and its reality—or, as they say, between the *Innenwelt* and the *Umwelt*," in other words, a "relationship between the movements made in the image and the reflected environment, and between this virtual complex and the reality it duplicates—namely, the child's own body, and the persons and even things around him."[120] The king of the microscopic human population, Lavon, looks up at the sky and sees himself, locates himself within the world as reflected back to him: "Yet, as always, Lavon's bemused upward look gave him nothing but his own distorted, bobbling reflection, and a reflection of the plant on which he rested" (405).

This reflecting threshold establishes the parameters of reality: "We're . . . living in the real world. Everybody these days knows that there's no other world but this one" (412). The real world is the space of life, of presence, of being as such, and the outside of this world can only represent an absolute negativity, a "no place" (412), a "burning void . . . where no life should go" (414), the absent space of death itself: "It had always been assumed that the plants [growing up from the bottom] died where they touched the sky. For the most part, they did, for frequently the dead extension could be seen, leached and yellow, the boxes of its component cells empty, floating embedded in the perfect mirror. But some were simply chopped off" (407). The "perfect mirror" marks the end of the universe, the boundary between life and death, the razor edge between presence and absence, "chopping off" anything that touches it, that tries to penetrate it, making it "impossible to pierce the sky" (406).

Despite the force of received belief, however, there exist traces of a world beyond. The original explorers from Earth left a series of nanolithographic tablets recounting for their microscopic descendants the history of how they came to be and the existence of a macroscale world—a more properly human-sized world—outside the confines of the pool. Hoping the tablets would serve to guide the microscopic people to "win their way back to the community of men" (398), the explorers were nevertheless aware that these tablets might "saddle their [descendants'] early history with a gods-and-demons mythology they'd be better off without" (398). For in announcing the existence of another world outside what can be seen and known by microscopic human beings, the tablets perform a mythological, Mosaic, biblical function. And indeed, over time and with the loss of some of the tablets, this mass of writings came to mystify and obscure. Ironically, the lithographs have the effect of reinforcing the small world as reality itself and its beyond as unthinkable: "There had never been anything in the plates but things best left unthought" (405). The unthought, the impossible, that which resists representation and imagination, is the real itself, the residue that remains after symbolization. It is also the space of death, the end and the origin of being, and therefore also the space of divinity, of the unknowable beyond of scriptural representation. As Derrida has shown, the history of logocentrism has always comprehended writing within a theological register, such that the transcendental signified implied by phonetic writing—especially biblical writing, or the book as such—is made by definition impossible, unthinkable, ungraspable, outside living speech, outside the signifier, outside the world: a presence beyond presence, yet inaccessible.[121] The nanolithographic plates both disclose and bar the outer void, the nothing-but-unthinkable of the text, sublimated then as the realm of the dead, the realm of the gods. Which is why Lacan has said, "*The gods belong to the field of the real.*"[122]

But in bringing the ontotheological implications of graphic writing, of mirrors, of thought and vision, to the surface of representation, the narrative of "Surface Tension" works to release the question of the beyond from humanist metaphysics. The story depicts the impossible made possible, and like nanovision more broadly, it exfolds the macroscopic outside to the molecular inside by expanding the limits of fabrication into endless frontiers of possibility, leaving no space for a transcendent sublime to remain hidden or veiled. Lavon, though he disbelieves that another world could exist beyond what he can see as the boundaries of his own world, nevertheless begins to think otherwise: "Perhaps this ['chopping off' of a plant at the surface of the sky] was only an illusion, and instead it [the plant] soared indefinitely into some other place—some place where men might once have been born, and might still live . . ." (407). He climbs one of the plants that grows upward to the face of the mirror—and perhaps through it: "Why could the sky not be passed, after all?" (407). Like the intrepid hero of "Jack and the Beanstalk," Lavon ascends from the microscopic human world toward the sublime world of giants.

As he approaches the limits of his world, the "surface skin" of the sky (407), he finds himself inside a protective mirrored sheath: "All around the bole of the water plant, the steel surface of the sky curved upward, making a kind of sheath. He found that he could insert his hand into it—there was almost enough space to admit his head as well. Clinging closely to the bole, he looked up into the inside of the sheath, probing with his injured hand" (408). Lavon tries to penetrate into the other world while remaining enclosed within a reflecting skin, typical of the humanist drive to extend the body into other space while remaining unaffected, protected in its own imaginary field: the centered subject who looks outward across visual distance, or extends an appendage, or sends technological prostheses, surveillance equipment, exploratory satellites, out into space—and therefore never escapes the mirrored sheath of the imaginary. Lavon comes closer to the surface skin of the sky: "The glare was blinding" (408).

But it is possible to see otherwise, and Lavon's extended hand, his tactile "probing" within this sheath of sky, functions like a scanning probe microscope, for which, as the creators of the nanoManipulator have written, "haptic feedback guides the progress of an experiment (you can't see what you're doing, but you can feel what you're doing)."[123] So Lavon "touched the sky with one hand" (407). And it is this extension of haptic vision that enables him to surpass the imposed limits of his reality, his "native universe" (408). Indeed, Lavon's probing of the surface between scales, his finger becoming an interface with the world beyond perceptibility like the nanofinger of the STM, sucks him into the sublime:

> There was a kind of soundless explosion. His whole wrist was suddenly encircled in an intense, impersonal grip, as if it were being cut in two. In blind astonishment, he lunged upward.
> The ring of pain traveled smoothly down his upflung arm as he rose, was suddenly around his shoulders and chest. Another lunge and his knees were being squeezed in the circular vine. (408)

In this highly sexualized description, the surface of the imaginary becomes a *sphincter*—not a barrier, not a limit, but a contractile stoma, a gripping and encircling "ring of pain" that squeezes and funnels him into the macroscale world. The human tunnels into the sublime real—"I've been beyond the sky" (410)—and promptly explodes:

> The water came streaming out of his body, from his mouth, his nostrils, the spiracles in his sides, spurting in tangible jets. An intense and fiery itching crawled over the entire surface of his body. At each spasm, long knives ran into him, and from a great distance he heard more water being expelled from his book-lungs in an obscene, frothy spurting. (408)

This "obscene, frothy spurting" from the "entire surface of the body," this whole-body orgasm, is not localized to some secure phallic object, some finite point of connectivity, detachable from the centered and protected subject, but instead exemplifies what Lacan has called the "jouissance of the body" *released* from the cohesion of the imaginary, no longer constrained by the symbolic structures of traditional thought. Lavon's explosive discharge in every direction, from every orifice, marks his accession to what Lacan hypothesized as a possibility for a "jouissance beyond the phallus," a radical dissolution of the centered subject into its multiplicities of sensation and experientiality.[124] Thus in tunneling across the mirror, through the looking glass, escaping the symbolic limits of fabrication, Lavon opens his organic being to the touch of the real.[125] Yet in having this shock of pleasure and pain, Lavon falls: "With a final convulsion, he kicked himself away from the splintery bole, and fell. . . . Sprawling and tumbling grotesquely, he drifted, down and down and down, toward the bottom" (408).

Like Icarus, whose hubristic transgression of human limits, whose approach to the boundary separating men from gods, is punished by burning and a mortal return to the earth, Lavon too is scorched, suffering "acute desiccation and third degree sunburn" (409), and plummets back into his "sunken universe" (409). But where the mythological fall of Icarus—like the fall of Satan, or Adam and Eve—serves to illustrate laws imposed on human possibility, "Surface Tension" employs the fall of Lavon precisely to overturn the metaphysical implications of "transgression."[126] The fall of Lavon is less a punishment for his daring against the limits of his world, his attempt to achieve the impossible, than an event that releases him from the perceptual boundaries of the closed world and the visual imaginary. For Lavon does not die like his counterpart Icarus but instead, in falling, returns to an infantile state: he sporulates.

> The spore-forming glands had at once begun to function. . . . The healing amniotic fluid generated by the spore-forming glands, after the transparent amber sphere had enclosed him, offered Lavon his only chance. . . . The sleeping figure of Lavon, head bowed, knees drawn up to its chest, revolved with an absurd solemnity. (409)

He is made embryonic once again, but now possessed of the knowledge that there does exist a real space beyond the mirror. Lavon becomes a fragmented entity *outside* and *anterior to* the mirror stage, a primordial beforeness produced by a dramatic reversal of that imaginary humanist drama, in Lacan's words, "whose internal pressure pushes precipitously from insufficiency to anticipation—and, for the subject caught up in the lure of spatial identification, turns out fantasies that proceed from a fragmented image of the body to . . . an 'orthopedic' form of its totality."[127] As a body in pieces, as the Lacanian *corps morcelé*, Lavon is interdigitated with the real and able to rebuild himself "from the bottom up." A demolished man, Lavon can start over from the fragments and redesign his subjectivity differently (just as nanoscience imagines tearing everything apart and reshaping it otherwise). Lavon's skin—his perception of bodily surface, his "armor of an alienating identity that [marks] his entire mental development with its rigid structure"[128]—is literally peeled away from his encounter with the ferocious real ("Lavon's body seemed rapidly to be shedding its skin, in long strips and patches" [409]), and he is born again without perception of *limitation*. He is, as Lacan's embryo, open to anything.

The opening of Lavon's body, even at its surface tissues, to newness and alterity—"His whole body shone with a strange pink newness" (410)—suggests the removal of the restraints separating inside from outside, the imaginary armor between the Innenwelt and the Umwelt. If Lavon's old skin performed a physical and imaginary enclosure against the environment, against the outside as such—an emblem of "the old fear of the outside" (415)—his "strange pink newness" allows him to interface with the world without mediation, in direct and intimate touch. His raw, peeled body meets the "raw landscape" of the world without a protective sheath, flesh to flesh (420). This stripping of skin as the stripping of barriers between imaginary and real anticipates, then, Lavon's subsequent voyage to the world outside the puddle in a ship designed to withstand the molecular forces operating at the surface of the sky. In a repetition of Lavon's previous speculation into the beyond—like some Fort/Da compulsion to repeat—the members of the expedition encounter the sky as "nothing but a thin, resistant skin of water coating the top of the ship" (415), a "glassy, colorful skin" that separates the small world from the larger one (416), a membrane beyond which they cannot see; refracting sunlight from the other world—a light that "seared [Lavon's] eyeballs" (416)—the "skin of water" is once again a literal "blinding barrier" (415).

Ocular vision fails on this side of the reflecting skin, but rather than finding themselves thus "forced to stop here on the threshold of infinity" (415), the microcosmic explorers extend themselves beyond the skin and thereby open toward the infinite space of the outside by physically probing through the barrier, by touching the other side: "I can see the top of the sky! From the *other* side, from the top side! It's like a big flat sheet of metal. We're going away from it. We're above the sky, Lavon, we're above the sky!" (416). In tunneling beyond the limits of perception, the explorers can see once again, but now "from the other side." They can see the outside as such: "Lavon saw Space" (416). Like Lavon's stripped skin, the membrane between worlds peels back, and the violated skin, no longer a "blinding barrier," proves to be an interface: "We are on a space-water interface, where the surface tension is very high" (417). Travel between these worlds does not involve simply traversing distance but demands an opening of one world to the other, a peeling of skin, a shredding of barriers—and at the bleeding graft between worlds, in this haptic space where one opens to the other in intimate touch, the difference between them is not one of distance, or even of scales, but mere surface tension.

The success of the expedition from one side of the perceptual surface to the other, the discovery that the skin is less a barrier than an interface, or a sphincter, becomes a valorization of the kind of thinking beyond limits characteristic of nanovision, characteristic of speculative science and speculative fiction as such: "what Man can dream, Man can do" (425). The double movement of specularity and speculation has made the impossible possible. The limit of fabrication, the space between representing and intervening, has vanished: "There is nothing that knowledge cannot do. With it, men . . . have crossed . . . have crossed space . . ." (425; ellipses in original). The crossing of space staged in this story means not only a travel across distance (embodied as a barrier or gap) but an abolition of distance entirely, for space has been "crossed" in the sense of having been "thwarted" and "disobeyed"—or even "crossed out," put "under erasure."

The "crossing (out) of space" here exemplifies a kind of posthuman logic in which the erasure of limits allows one world to inhabit the other, and challenges the humanist drive to mastery whereby the protected, ocularcentrist self conquers other worlds across distance. When there is no distance, when the inside is opened up to the immensity of other space and the radical endlessness of overlapping frontiers, there is no longer any god's-eye view from which the human can look out on the other world with the security of mastery, because there is no longer any space left for the ineffable transcendental, for the absolute sublime or divine. As Lavon puts it: "We're beyond the sky, all right. But we're not gods" (424).

Or as Shar, the philosopher-scientist on the expedition to the macrocosm, says: "We're too small. Lavon, the ancients warned us of the immensity of space, but even when you see it, it's impossible to grasp. And all those stars—can they mean what I think they mean? It's beyond thought, beyond belief" (422). The explorers recognize that the immensity of real space cannot be contained by the human symbolic field or human vision; it is always "beyond thought, beyond belief." But inasmuch as they themselves have already crossed beyond the "threshold of infinity" and have emerged into the impossible beyond, "even in seeing it, it's impossible to grasp." They have touched the real, but they cannot "grasp" it. They have merged with the outside, they have revisioned an imaginary limit as instead an interface with alterity, they have come to live in the strange new world, but this inhabitation merely proves to them that they are indeed "too small" to domesticate it.

Even attempting to encapsulate the immensity of space into a "world picture" leaves them only yet again with the endlessness of infinity, the failure of a totalizing world picture that can only locate itself within an already transcendent multiplicity. Shar says, "I'm beginning to get a picture of the way the universe is made, I think. Evidently our world is a sort of cup in the bottom of this huge one. This world also has a sky of its own; perhaps it, too, is only a cup in the bottom of a still huger world, and so on and on without end" (417). In having entered the sublime real, in having taken up residence within the infinite real space "beyond thought, beyond belief," the explorers can no longer imagine the possibility of an "end" that is not *already exceeded*, and therefore they can no longer imagine totality.

With the collapse of distance between worlds enabled by nanovisual perception, the beyond is not beyond; it is already enfolded into the endless frontiers of an infinite multiverse. Depriving human perception of a privileged spectatorial position from which to take control of its imagined environment, nanovision problematizes the relationship between the observer and the observed, the human and the extrahuman, the macroscale and the nanoscale. Even as Lavon pops out into the new world and "sees Space," simultaneously the "insane brilliance of empty space looked Lavon full in the face" (415). Space sees itself being seen. This is the moment when the full implications of the logic of nanovision appear: when distance vanishes in the haptic space of closeness and molecular intimacy, the observer becomes the observed, and vice versa. When there is no distance, the human touches, and the real touches back.

Thus the logic of nanovision that seems at one moment to support human aspirations for the "absolute control" of the nanoworld simultaneously confronts them: nanovision is in tension with itself, the surface differences between seeing "too small" to capture the world and seeing "too big" to relinquish it becoming simply a matter of perspective, a question of from which side of the looking glass one peeps.

For at the same time, Blish's story shows how easily the posthuman imagination of nanovision comes to be reasserted in a worldview of human mastery. Only moments after Shar's "vision of space" suggests the impossibility of domestication (417), Lavon remarks: "I can see that this world needs a little taming" (423). The colonialist imperative of the visual ("I can see") takes over here, at the very moment of touching the real, in the same fashion as it does in so many areas of nanoscience. But this scopic drive for mastery of the new world can only exist with the reappearance of *distance*. "Surface Tension" ends on a note of ambiguity to express precisely that the escape from humanist limitations made possible in the collapse of distance, the "crossing" of space, becomes yet again an imprisonment in the human-ist psychic apparatus when distance is allowed to reassert itself:

> "We have crossed space," Lavon repeated softly.
> Shar's voice came to him across a great distance. The young-old man was whispering: "But *have* we?"
> "As far as I'm concerned, yes," said Lavon. (425)

There the story ends, and the reappearance of distance at the end of the fiction—"Shar's voice came to him across a great distance"—occurs in con-nection with the reassertion of the authoritarian humanist subject, the "I" who proclaims the limits of reality, the conditions of possibility, the ontology of the world and the finitude of representation. Let us not forget that Lavon, the speaking subject, the "I" who states what has occurred and who declares an end to the story—a limit to fabrication—is also the king, who carries the same name as all Lavon-kings before him. "Lavon" is, indeed, the "name of the father," and the symbolic field that determines what is imaginable is reformed by the central subjective "I" who sees and states that what *it* sees is what *is* as such: "As far as *I'm* concerned, yes."

But why does the story conclude with this ambiguous questioning of whether humans have crossed space or not? It would seem to do so in order to take account of the multiple meanings of "crossing space" at play here. If only moments earlier the "crossing of space" meant a collapse of distance, now, within the colonialist mentality that has come upon the explorers from the microcosm (or at least upon Lavon, the king, if not Shar, the scientist, who now finds himself at an irresolvable distance from Lavon), the crossing of space takes on an entirely new meaning.

Rather than being diminished, space has now been allowed to constitute a fundamental and absolute distance between subject and object—that dis-tance required for mastery—and therefore space has, once again, become sublime. In other words, "men have crossed space" in the sense that space has been *crucified*, and thereby *deified*. Space has once again come to be the impossible, transcendental, sublime real, the very space of divinity; and *nailed up on the cross*, "real space" is sacrificed to the limitations of the sym-bolic. It recedes, once again, into an impossible beyond. Humans may have crossed space in numerous ways in this early story of nanovision, but the fiction ends with the strong sense of missed opportunity, for the opening of a way of seeing otherwise comes out, in coming to an end, to be once again yet another excuse for colonialist expansion and a reinscription of the metaphysical structures that permit domination and exploitation of the other.

INTERFACIAL ETHICS/MOLECULAR EROTICS

While shrinking the physical distance between here and there, providing access to new landscapes and new properties, nanovision constantly reproduces the metaphysical, or ontotheological, distance of mastery—the distance instantiated in the nanodiscursive tropes of absolute control, small worlds, territorial corrals, and microcosmic gods. Indeed, Theodore Sturgeon's "Microcosmic God" (1941)—another powerful coordinate in the literary prehistory of nanotechnology—suggests that the distance of mastery increases even as perceptual distance between the human and "the outer limits" decreases. In his remote island laboratory, the polymath scientist Mr. Kidder evolves a population of microcosmic engineers, minute beings called Neoterics, who (like Drexlerian assemblers) can fabricate any material or technology from the bottom up. He instills a fearful obedience in the Neoterics by intervening in their world with destructive force. And yet, even in having gained complete control by extending his physical reach into their world, Kidder can only see them across the *telescopic* distance of power: "Such are the results of complete domination. . . . Crouched in the upper room [of the lab], going from telescope to telescope, running off sloweddown films from his high-speed cameras, he found himself possessed of . . . a new world, to which he was god."[129]

The inaccessible space of divinity, the impossible beyond of infinite distance, appears even as an effect of traveling to the limits of the perceptible world. That the microcosmic god does not see his subjects in real time and views them only abstractly, as from the absolute outside, through sloweddown films of high-speed telephoto images, in itself shows that even when *crossing* distance one way, nanovision *crosses* distance another way: a double movement that reaches out to the real only to destroy its possibility once again. We can understand, then, precisely why Emmanuel Levinas has insisted on an *irreducible* distance from the other in the face-to-face as the condition of ethical relations.

Nevertheless, even in crossing the space between worlds, nanovision also crosses *out* that space, and this speculation beyond enables an unprecedented *interface* between self and other, between our world and the nano-world, and between present and future that ultimately deconstructs the very possibility of mastery. Even at the moment that nanovision lets us set up empires and futures markets in molecular space, molecular modes of being encroach into us, ambiguating the formal boundaries of the humanist sub-ject and its temporality.

This recognition of bidirectional flow across the interface motivates Stephen Baxter's story "The Logic Pool" (1994), which neatly rewrites "Microcosmic God" by way of "Surface Tension."[130] "The Logic Pool" re-counts a nanovisual system in which the eccentric scientist Marsden creates sentient nanoscale logic structures, "nanobees," with whom he communi-cates through a neural implant in his head. The nanobees, perceiving the domineering Marsden as "some huge god outside their logic pool" (169), strike out with "murderous fury" in an act of reverse colonization (169), tun-neling across the interface—"They got through the interface to Marsden's corpus callosum" (171)—to physically invade the domineering creator-being outside their world. One nanobee rages against the surface of the interface with Marsden: "And, for a precious instant, he reached *beyond the Sky*, and into something warm, yielding, weak" (155). The inhabitants of the nano-world thwart the distance of mastery, recrossing it to cross it out, and in that crossing find themselves without limits: "The Pool beyond the Sky was limitless. [They] could grow forever, unbounded" (173). That the man who would be king, or microcosmic god, is destroyed by this incursion of the nanoworld perhaps goes without saying, for the nanovisual interface itself undermines the proper boundaries of closed subjectivity, the proper bound-aries of the human as such. When the nanoworld interfaces with the human world, invading us even as it has been invaded, all that remains of the sub-lime space of divinity is a pathetic "faint smell of *human*, a stale, vaguely unwashed, laundry smell" (158).

If we are moving into the nanoworld, the nanoworld is moving into us, and if this demolition of human boundaries signals the end of easy mastery and petty gods, it also signals the opening of a way of interrelating with forms of molecular being that are not predicated on the logic of control. In-deed, it is nanovision's ability not only to see up to the limits of fabrication, to cross symbolic distance, but also to even go beyond those limits, to extend outward to radical otherness, to reach out from the possible into the impos-sible, to solicit the outside as such across an open tunnel and let exteriority inside in a disruptive and transformative way—in other words, its drive to see things only impossible until they are not—that creates the conditions for a posthuman *ethics of the interface.*

While nanovision clearly does not in itself subvert totalitarian forms of violence toward the other, it does introduce another form of violence, an interfacial violence—the peeling of skin between self and other—that marks a fundamental intimacy with the other. This kind of masochistic intimacy enables the self to become other, and thereby a posthuman ethos of self-alienation is made possible by thinking nanotechnologically. The interdigital stripping and renovation of the membranes between interiority and exteriority, the fingering, probing, and touching characteristic of nano-vision's specula and speculations, engineer an equivocal splice site where self and other *correspond*, communicate, and pass into each other, where one does not consume the other but where otherwise autonomous entities bleed and flow together. Nanovision discovers its own erotics.

The ethics of the interface, then, would appear as what we might call *molecular erotics.* Imagine lovers being able to caress each other's very molecules! Or even able to merge, gaseously, into each other's very space of being. To make contact with formerly unknown erogenous zones of the world itself, the materiality of the world in its molecular flux. The intimacy of the nanoprobe, the subjective tunneling across the interface into the ob-jective, the merger of self and other at the place of the sensitive fingertip, the touch of raw flesh to raw landscape . . . through the haptic mediations of nanovision, we enter into an erotic relation with the nanoworld and with the innermost spaces of our bodies, the innermost spaces of our world, even as the nanoworld enters us and overflows the outermost spaces of our known frontiers. Though of course the nanoworld has always been there, we only now begin to recognize it at the unbounding of our world, the unbounding that occurs when we touch the other and the other touches inside us. The haptic visualization practices of nanotechnology, as we have seen, mediate an unprecedented intimacy with matter. Don Eigler recounts his own erotic sensations of touching and observing formerly imaginary nanoscale objects and quantum effects through the probe-microscopic interface: "We knew of them in a purely cerebral way. But here they are, alive to our eyes and responsive to our hands . . . quantum mechanics made visceral! . . . [They] evoke a delectable intimacy between us and the quantum world."[131] And it is exactly this level of visceral intimacy, this delectable erotics of touch, that ensures our recognition of the molecular other *as* other, that provides the basis for an ethical relationship with nanoscale alterity transverse to the world picture of absolute control. As Binnig and Rohrer write:

Since the advent of local-probe methods, atoms, molecules, and other nanometer-sized objects are no longer "untouchables." They forsook their anonymity as indistinguishable members of a statistical ensemble and became individuals. We have established a casual relationship with them, and quite generally with the nanometer scale. Casual, however, does not mean easy. They are fragile individuals. . . . Interfacing them to the nanoscopic, microscopic, and macroscopic worlds in a controlled way is one of the challenges of nanotechnology. Imaging them or communicating with them is the easiest of these tasks, although not always trivial.[132]

Even as we touch the nanoworld at the interface, other faces appear and solicit our caress. Nanoscale beings emerge from the "anonymity" of the mass or the crowd, away from the easily victimized "statistical ensemble," and become "individuals" with whom we develop a "casual relationship," a friendship and an intimacy that respects their "fragility." We touch them gently with the local probe (unlike the destructive violence of an electron beam), and they respond. We "communicate" with them. They become recognized as others beyond ourselves and beyond our world in coming to "interface," and while nanotechnology works to make them interface, to "control" them into interfacing with our world, the very act of the interface in itself, crossing the distance between worlds by touching, enables us to see, or at least to "image," the inhabitants of those worlds as worthy of our respect, as needing our care toward their "fragile" bodies. Our awareness of this tactile relationship, this erotic correspondence, is not to be taken as "easy" or "trivial." They are no longer "untouchables," not because we have violated some prohibition on their virginal purity, but because our perception of prohibition—whatever "law," physical or symbolic, that would have made them untouchable—vanishes once we enter into ethical relations with them, once we enter into the complexities, the nontrivial conversations and negotiations, of what here in Binnig and Rohrer's account seems very much like the condition of *espousal*. For have we not just read what, in a different context, would be the depiction of a marriage, with all its undertones of courtship, commitment, sexual objectification, yet above all, love? But we are getting ahead of ourselves . . .

Enabling this intimate interface with the other despite—or perhaps, perversely, even because of—its logic of control, nanovision would seem therefore to approximate the kind of "eschatological vision" or "transcendent vision" theorized by Levinas as the interfacial encounter with alterity, where the gleam of infinity (i.e., radical otherness) overflowing visible totality (i.e., the imaginary map or world picture) appears in our communications and correspondences with the other as other, in the very face of the other that resists assimilation. The transcendent vision institutes "a relationship with *a surplus always exterior to the totality*, as though the objective totality did not fill out the true measure of being. . . . The experience of morality does not proceed from this vision—it *consummates* this vision; ethics is an optics. But it is a 'vision' without image, bereft of the synoptic and totalizing objectifying virtues of vision, a relation or an intentionality of a wholly different type."[133] In its speculations beyond the limits of fabrication, nanovision bodies forth the possibility of a "wholly different" ethical vision that does not totalize or capture with its gaze but instead enters into a "casual relation" with the real exterior to the imageable. With "the touch that makes the difference," the molecular erotics of nanovision facilitates the interpenetration, or interdigitation, of worlds. It fundamentally disallows the viewing subject to remain autonomous, but rather insists that the subject be transformed by its fusion with the object, that the nanoworld be recognized as dwelling within us as much as we might dwell within the nanoworld.

Nanovision therefore enables a new, fundamental intimacy with the other even while promoting the absolute control of matter; even within the framework of humanist colonialism, it begins to erect a different framework of posthuman ethics and responsible molecular erotics. Evidently there are options here, opportunities for other futures, alternate futures, which may or may not be achievable. It remains to be seen how nanovision will shape itself, or be shaped, as the future unfolds. There is still time to decide, to intervene, to engineer the future. For while the nanofuture perhaps shows all signs of having already arrived—and while, as we will see in the next chapter, a certain resistant humanism in the political discourse on nanotechnology seems to diminish the radically alienating possibilities of nanovision—the future still remains saturated with possibility.

3

THE HORRORS OF GOO:

*Molecular Abjection and the Domestication
of Nanotechnology*

> If we can visualize the problems and begin to think
> about the impact of the nanotechnology revolution,
> we can begin to plan for the coming storm.
> —John G. Cramer, "Nanotechnology: The Coming Storm"
>
> There's a storm coming.
> —*The Terminator*

Must we resist? And, if so, nanotechnology?

After all, we find resistances even within nano itself. The very resistances that have released their surface tensions and given way to concepts like "science (fiction)" or "molecular erotics," the motions of resistance at the skin between possible worlds, are themselves symptomatic of nanotechnological ways of seeing. By virtue of the membranes of resistance within the technoscapes and dreamscapes, that is, by virtue of their terminal instability, nanovision encounters and speculates on the bleeding interfaces between our world and the nanoworld, between the human present and the posthuman future, between fantasies . . . and nightmares.

NANOCULTURE AND ITS DISCONTENTS

So it should come as no surprise that nanotechnology's spectacular rise to prominence in the early years of the twenty-first century has generated in tandem a flood of increasing concerns over the risks, both mundane and extreme, associated with the future of this emergent technoscience.[1] As early as 1980, Drexler and others involved in prophetic discussions of the nanofuture had begun to consider threats that might accompany molecular manufacturing capabilities, eventually concluding that the nanofuture would require endless societal diligence to ensure that nanodreams would not give way to calamity: "The choices we make in the coming years will shape a future that stretches beyond our imagining, a future full of danger, yet full of promise."[2] More recently, apprehensive voices and emissaries of discontent in the dawning era of nano have drawn public attention to a panoply of anticipated dangers: for example, the toxicity of nanoparticles and environmental pollutants from nanofactories; copious opportunities for nanoterrorism or nanowarfare; potentially dramatic economic imbalances of globalizing nano-industries; a host of ethical risks from increased medical surveillance to failures of democracy in the nanofuture; and even the shocking possibility that autonomous nanobots might rise up in rebellion and wreak havoc on the world.[3]

Alarmed authorities as diverse as the British Royal Society and the Royal Academy of Engineering, Greenpeace, the Center for Responsible Nanotechnology, Swiss Reinsurance Company, the Foresight Institute, and several individual nanoscientists have published extensive cautionary reports recommending measures to protect ourselves physically and socially from the onrushing hazards posed, ironically, by our "ultimate control of matter."[4] Some, like the Canadian ETC Group, have called for a moratorium on nanoresearch until its dangers have been more adequately assessed.[5] Others, traumatized by posthuman nanovisions, have advocated instead complete "relinquishment." Bill McKibben, for one, has written that "anything with the power to make us 'posthuman' should be watched with a beady eye; each incremental advance should be presumed dangerous until proven otherwise," and perhaps before nanotech and allied sciences progress any further, we should all just say, "Enough."[6] Even more apocalyptically, Bill Joy, cofounder and former chief scientist of Sun Microsystems, sent shock waves through the digerati in 2000 when he wrote that nanotech bodes such inherent catastrophe for our planet and our continued existence that "the future doesn't need us." For if ongoing advances in nanotech, genetic engineering, and robotics provide "tools which will enable the construction of the technology that may replace our species," and if "our own extinction is a likely, or even possible, outcome of our technological development," we should wisely, and before it is too late, give up our claims on the nanofuture.[7]

The number of anxious citizens and vocal critics taking nanotechnology seriously as "real science" has multiplied significantly over the last decade, and as Kate Marshall writes, within the field of nanodiscourse, where the present constructs itself in reflection of an already inevitable future, "risk as such emerges as the most powerful force in determining the nature of the nanotechnology represented in the popular forum."[8]

While nanotechnology embodies a scientific fantasy of complete control over the material world, its capacity to mechanically engineer nature becomes simultaneously the source of its greatest dangers. Many perceived risks of nano relate to unanticipated consequences of designing nanomaterials for self-organization and problems of emergent properties in complex systems, especially those capable of autosynthesis and self-replication. Of course, as George Whitesides has written, engineering nanosystems for self-organization makes molecular manufacturing more feasible:

> *The smaller, more complex machines of the future cannot be built with current methods: they must almost make themselves.* . . . A self-assembling process is one in which humans are *not* actively involved, in which atoms, molecules, aggregates of molecules and components arrange themselves into ordered, functioning entities without human intervention. . . . The approach could have many advantages. It would allow the fabrication of materials with novel properties. It would eliminate the error and expense introduced by human labor. And the minute machines of the future envisioned by . . . nanotechnology would almost certainly need to be constructed by self-assembly methods.[9]

But at the same time, systems endowed with the ability to self-organize, to self-direct, to self-determine, evoke one of the most entrenched horror-fantasies of modernity: the possibility that our machines will escape our mastery and run amok, that our tools will revolt, destroy us or enslave us, that our technology will go "out of control."[10] When Heidegger diagnosed the essence of modern technology as the "enframing" of our world and our selves by technology—the same determinism we see everywhere in nanonarratives of the "already inevitable"—he called it "the monstrousness that reigns" in the present age.[11] Current fears of nanotech escaping control doubtlessly emerge at least in part from the prevalent attitude that its development is already mapped out—that the nanofuture is already here, or that the nanoworld is truly small, after all—and that we are the mere medium of its delivery. In other words, that nanotechnology has already exceeded its human creators. Older than the mythology of Mary Shelley's *Frankenstein* (1818), the anticipatory fear of technology running amok, along with the recursive fear that it might not be we who master technology but instead technology who *already* masters us—both monstrous violations of "natural order"—has been endemic to conceptions of technical change in modern history.[12] So if one of the fundamental feasibility requirements of molecular manufacturing involves imbuing nanosystems with massively self-organizing or self-assembling capabilities—technical operations whose cumulative scaling effects so far remain in the realm of absolute uncertainty—then it would seem that the advancement of nanoscience gives force to some of the greatest anxieties of technoculture.[13]

The main scare scenario most frequently associated with nanotechnology is the notorious "gray goo problem." Imagine nanomachines frantically reproducing themselves and spreading like a raging flood across the globe, eating everything in their path and transforming it into copies of themselves, eventually converting the whole planet into amorphous and chaotic flows of disorganized matter, colorless and lifeless goo. The nanoscientist Robert Freitas writes, "Perhaps the earliest-recognized and best-known danger of molecular nanotechnology is the risk that self-replicating nanorobots capable of functioning autonomously in the natural environment could quickly convert that natural environment (e.g., 'biomass') into replicas of themselves (e.g., 'nanomass') on a global basis, a scenario usually referred to as the 'gray goo problem' but perhaps more properly termed 'global ecophagy.'"[14] The danger of gray goo is clear: nanotechnology could devour the world. Drexler addresses the problem in *Engines of Creation*:

> They [self-organizing nanobots] could spread like blowing pollen, replicate swiftly, and reduce the biosphere to dust in a matter of days. Dangerous replicators could easily be too tough, small, and rapidly spreading to stop—at least if we make no preparation. We have trouble enough controlling viruses and fruit flies.
>
> Among the cognoscenti of nanotechnology, this threat has become known as the "gray goo problem." Though masses of uncontrolled replicators need not be gray or gooey, the term "gray goo" emphasizes that replicators able to obliterate life might be less inspiring than a single species of crabgrass. They might be "superior" in an evolutionary sense, but this need not make them valuable. . . .
>
> The gray goo threat makes one thing perfectly clear: We cannot afford certain kinds of accidents with replicating assemblers.[15]

Gray goo, whether produced by a laboratory accident or by malicious intent, could bring planetwide cataclysm. But the scenario is one that many critics have deemed to be "impossible" or "far-fetched." In fact, several advocates of nanotechnology have displaced the threat of gray goo from the register of the real into the imaginary social field—a danger, that is, to the cultural prestige and progress of nano as a technoscientific practice.

Richard Smalley, especially, worried for many years before his untimely death in 2005 that talk of gray goo would lead to widespread public misunderstanding of nanoscience. Hoping to protect the field's image, he argued in 2001 that fears about nano are woefully misguided:

> The principal fear is that it may be possible to create a new life form, a self-replicating nanoscale robot, a "nanobot." Microscopic in size, yet able to be programmed to make not only another copy of itself, but virtually anything else that can be imagined, these nanobots are both enabling fantasy and dark nightmare in the popularized conception of nanotechnology. They would enable the general transformation of software into atomic reality. For fundamental reasons I am convinced these nanobots are an impossible, childish fantasy. . . . We should not let this fuzzy-minded nightmare dream scare us away from nanotechnology. Nanobots are not real.[16]

Childish fantasy. Impossible. Unreal. The same words so frequently used as weapons against nanotechnology now come to its defense. Even from within. Whether or not Smalley is ultimately right about the impossibility of nanobots—though it seems to be the case that he and Drexler represent such radically incommensurable scientific cultures as to render this question itself undecidable under the given terms of their debate—such concerns have galvanized the technoscape.[17] The likelihood that this "fuzzy-minded nightmare" might indeed scare us away from the nanofuture prompted the U.S. National Nanotechnology Initiative to proclaim on its website in 2003 that while the dangers of self-replicating nanotechnologies appear potent, "Such creatures do not exist and many scientists believe they never will. . . . Today—and as far as scientists can see—anything resembling nanobots remains in the realm of science fiction."[18] Mihail Roco, in his role as director of the NNI, has similarly said that "the theory of self-replicating nanobots taking over the world" is "very speculative, more like science fiction."[19] This move to dissociate catastrophic risks from nanotechnology proper by dumping them into the domain of science fiction is, as we have seen, representative of the most common strategy that nano has employed throughout its history to legitimate itself as real science. But it also suggests the extent to which promoters believe that perceptions of danger could damage the field in the same way that the imagined risks—physical, moral, or otherwise—associated with human cloning, stem cell research, genetically modified foods, and so forth, have led to legal prohibitions or public distrust (with financial consequences) of certain forms of scientific inquiry.[20]

So even if some dangers of nanotechnology are not "realistic," nano-rhetoricians still recognize them as dangers—dangers to nanotechnology itself. As the chemist Emmanuelle Schuler writes, "The failure to understand or acknowledge how non-technical persons perceive, assess and make decisions about risk may hamper the trajectory of nanotechnology as public policies and practices are adopted. In conclusion, perceived risks are real. Perceived risks may very well constitute the tipping point that will decide whether nanotechnology succeeds."[21] Burson-Marsteller, one of the world's largest public relations companies, has also advised the nanotech community: "Although widespread discussion of the opportunities and risks associated with nanotechnology has not yet occurred, there is already a danger that public acceptance will not be good. . . . Nanotechnology is facing a challenge: dialogue with the public and information about the subject can be an effective way of dealing with emerging fears."[22]

The nanotechnology orthodoxy, therefore, has recently opened this dialogue, but largely to downplay the dangers of nanoscience—or indeed, to dismiss the problem by rejecting from the field those figures associated with apocalyptic visions.[23] For example, Mark Modzelewski, former director of the U.S. NanoBusiness Alliance, commented in 2003 on a notable silence regarding molecular manufacturing in U.S. nano legislation (namely, the Twenty-first Century Nanotechnology Research and Development Act): "There was no interest in the legitimate scientific community—and ultimately Congress—for playing with Drexler's futuristic sci-fi notions."[24] Apparently the "legitimate scientific community" and its business associates wanted nothing to do with the fantasies of "bloggers, Drexlerians, pseudo-pundits, panderers and other denizens of their mom's basements."[25] Smalley has likewise said that Drexler's entire theoretical project is a mere "bedtime story" good only for frightening children, and therefore better ignored. Smalley claimed in 2003 to have encountered a startling number of Texas schoolchildren who believed "that self-replicating nanobots were possible, and most were deeply worried about what would happen in their future as these nanobots spread around the world. I did what I could to allay their fears, but there is no question that many of these youngsters have been told a bedtime story that is deeply troubling." He accuses Drexler of pointless scaremongering:

> You and people around you have scared our children. I don't expect you to stop, but I hope others in the chemical community will join with me in turning on the light, and showing our children that, while our future in the real world will be challenging and there are real risks, there will be no such monster as the self-replicating mechanical nanobot of your dreams.[26]

Thus the "legitimate scientific community" rallies to implement damage control and uphold the innocence of nano, this technoscience that it has elsewhere dubbed the source of endless bounty and "the next industrial revolution,"[27] so unjustly maligned and misunderstood by its own "unhappy father."[28] Protecting the reputation of nano would then also seem to protect our children from monsters. Children be calm, children be unafraid: nanotechnology will not run amok and turn on us, it will not devour its legitimate creators, its legitimate fathers. Sleep peacefully, children, and be troubled no more by the horrific imagination of Dr. Drexler, the man who would be Frankenstein.

The gray goo problem, the monster of nightmare, the self-replicating nanobot fantasy, serves to focus a variety of fears about nanotechnology—everything from poisonous nanotubes to nanoweapons of mass destruction, every unanticipated consequence of engineering at the molecular scale that might indicate our technology has in some sense "turned against us." Evoking it has a strategic value whether one wishes to question the safety of nanotech research or affirm it. Because if gray goo can frighten children, it invites the entire nanotech generation to be skeptical and cautious about nanohype and the many promised wonders of the nanofuture. But at the same time, in focusing fears on an unlikely event in an equally distant future, gray goo also provides a convenient means of diverting public attention from other more plausible dangers. As Drexler and Phoenix have suggested, disproportionate fixation on gray goo—an extreme scenario of advanced nanotechnology that could be prevented with precaution and foresight[29]—distracts and detracts from more immediate risk scenarios:

> All risk of accidental runaway replication can be avoided, since efficient [molecular] manufacturing systems can be designed, built, and used without ever making a device with the complex additional capabilities

that a hypothetical "grey goo robot" would require. However, this does not mean that molecular nanotechnology is without risks. Problems including weapon systems, radical shifts of economic and political power, and aggregated environmental risks from novel products and large-scale production will require close attention and careful policy-making.[30]

The nanofuture is no easy utopia. But to the extent that some defenders of nanotech have metonymized the very thought of nanodanger in the red herring of the gray goo problem, pursuing an improbable straw man only to sacrifice it in the sphere of public discourse, the dystopian features of any imagined future recede into the background. When Smalley and others "turn on the light" against the unseen and "impossible" horrors of self-replicating machines—banishing gray goo to "mom's basement" and the underside of children's beds as an imaginary bogeyman, dismissing critical discourse on the dangers of molecular manufacturing as "not real" and "childish fantasy"—the future is made bright for nanoscience once again.

But if the horrors of goo are so palpable that they must be repetitively and ritualistically dismissed as pure fantasy to protect nanotechnology from condemnation, it begs the question of what kind of cultural work a fantasy like this is performing. For the fantasy is ridiculed even when presenting itself as nothing more, or less, than fantasy as such; for example, Michael Crichton's *Prey* (2003), a science fiction novel about nanogoo, continues to be pilloried by the nanoscience community apparently for the crime of pretending to be exactly what it is, namely, science fiction: "Don't let Crichton's *Prey* scare you—the science isn't real."[31] But this fantasy arose not from outside nanotechnology but from within: gray goo, the idea of self-replicating nanobots escaping our control, is a theoretical scenario foreseen by the speculations of nanovision, a risk perception made available only by thinking nanotechnologically in the first place. It is a science (fiction) narrative formed inside the field of science (fiction). So the question would then be why gray goo became the "earliest-recognized and best-known danger of molecular nanotechnology," capturing so much more attention than other possible dangers and becoming such a potent cause for alarm both for those who believe it "real" and for those who believe it "not real." Real and unreal, science and science fiction, gray goo is a liminal concept at the interface of our symbolic field and its beyond, unrealizable in the present except as extrapolation into the future, but whose extrapolation nevertheless affects decisions regarding nanotechnology even now. Like all nanonarratives, gray goo exemplifies the way in which science fiction produces or motivates science. Even if impossible, or, I should say, especially if impossible, gray goo does things.

Specifically, the gray goo scenario in itself bodies forth a paradox: at the peak of its mastery, in its most evident "proof" of complete control over the structure of matter—the proof of fabricating self-sufficient nanomachines—nanotechnology has already lost control. In gray goo, the ultimate control of matter has ineluctably produced the ultimate example of matter in rebellion. Victor Frankenstein discovered this paradox, as well:

> How can I describe my emotions at this catastrophe, or how delineate the wretch whom with such infinite pains and care I had endeavoured to form? His limbs were in proportion, and I had selected his features as beautiful. Beautiful!—Great God! His yellow skin scarcely covered the work of muscles and arteries beneath. . . . Unable to endure the aspect of the being I had created, I rushed out of the room, and continued a long time traversing my bed-chamber, unable to compose my mind to sleep.[32]

Thus appears the unexpected consequence that, even as the direct outcome of perfect control, perfect precision, our technology can literally *revolt*. Machinic resistances to our will, figured as revolutions or overturnings from below, appear identical to the visceral regurgitations of *disgust*.

Hence the resonant fixation on nanotechnology and "goo": goo, as disorganized material, embodies the very essence of lost control, the absence of any organizing principle, any sense of command. "Goo" is the sound a baby makes before it has mastered language; "goo" is the primordial ooze out of which life emerged, and to which life returns; "goo" is entropy in metonymy, it is chaos made flesh; "goo" is the loss of information, the spillage of meaning, the death of coherence, the end of mastery.

The horror of goo is the horror of abjection. As Julia Kristeva has written, the fluid substances of the inside of the body spilled outside as goo—excrement, filth, putrescence, snot, semen, menstrual blood, urine, vomit—threaten subjective coherence even as their abjection works to preserve it. These abject flows of refuse "*show me* what I permanently thrust aside in order to live. These body fluids, this defilement, this shit are what life withstands, hardly and with difficulty, on the part of death. There, I am at the border of my condition as a living being. My body extricates itself, as being alive, from that border. Such wastes drop so that I might live, until, from loss to loss, nothing remains in me and my entire body falls beyond the limit—

cadere, cadaver."[33] In other words, goo is revolting because it represents the death of self-mastery, or death as such, the fundamental dissolution of the human subject as contained within its skin and its imaginary boundaries.

Which is why, as Kristeva shows, the abject is fundamentally associated with the inside of the maternal body, with the fantasy of a devouring mother and that primordial *indivision* of one body from another, that time before the body separates from its prenatal environment, its enveloping womb, its enatic extension. The abject substance as the before and after of subjectivation, the outside of the self before birth and in death, threatens the borders that the human subject has worked to establish, threatens autonomy, and even as a site of desire for the outside, for that which would plug the gap or intractable wound of the subject's severance from the outside, it is already prohibited as symbolic limit, the end of life. The abject substance both entices and horrifies, and it flows forth as abjection precisely to defend subjective security. Overcoming the repulsive matter that disturbs system, order, and regulation—that is, reasserting mastery—entails thrusting the goo aside with repulsion, forcing it outside even as it comes forth from inside. Or, as Modzelewski and Smalley have suggested, we must confine goo to the darkness of "mom's basement," or under the bed, as the stuff of children's bedtime stories and the maternal imaginary, the stuff of nightmares . . . the stuff, indeed, of *nocturnal emissions*.

If the attempt to master goo, to constrain it within a specific register of dominance and restraint, would therefore be what we might specify as "insemination" (the funneling of goo into a rigid signifier of masculinist control over emissions and their spread, the canalization of fluidity as such), then the failure of insemination, the failure of the principle of mastery—let's go ahead and call it the phallus—the failure of the phallus would then be what Derrida has called "dissemination." As the movement and overflow of the semantic and seminal from within, the multiplication of essential meanings and vital fluids, "dissemination—which entails, entrains, 'inscribes,' and relaunches castration—can never become an originary, central, or ultimate signified, the place proper to truth. On the contrary, dissemination represents the affirmation of this nonorigin, the remarkable empty locus of a hundred blanks no meaning can be ascribed to."[34] Dissemination spills meanings and origins and seeds and fecund potentials in advance of, and far beyond, the constraining demands of any phallic function: goo, in other words, overflows as the abject substance that would signify in multiplicity, polysemically, more than anticipated or permitted by traditional channels of the humanist symbolic. Goo disseminates, it spreads and floods, it washes and strips away, and in so doing, it generates new sites of resistance against the unitary fantasy of insemination, or control.

The gray goo scenario records a subterranean dread of technocracy discovering that its mastery of nature, pushed to fundamental limits of fabrication, might entail the loss of mastery through the very control mechanisms it has erected: in other words, the effort to capture the symbolic locus of technical power—which we might shorthand as the "technoscientific phallus"—will enact its own failure. A failure only overcome, as it were, through the abjection of self-organizing molecular systems and other forms of nanotech which threaten to overflow and expose the fictionality of the "ultimate control of matter." *Molecular abjection* as such.

Molecular abjection, then, would constitute an essential response to matter that revolts the logic of control. Revolting not only because it escapes, but moreover because it externalizes nanotechnology's very desire for incorporation, its drive to absorb the world, to reshape the world in its own image, which returns now as a devouring force from the outside, as "something rejected from which one does not part. . . . Imaginary uncanniness and real threat, it beckons to us and ends up engulfing us."[35] Goo revolts as matter which had already been contained or consumed by nanoscience now irrupting as that which devours us. Goo incarnates the horror of being eaten by an abject substance we have in fact fabricated ourselves, the perverse desire for an imaginary ecophagy instantiated by nanoscience's dreams of global domination made a real threat to our world: "THEY call it 'global ecophagy.' That's 'eating the Earth' to you and me. Rumour has it that this is what replicating nanostructures might do, and according to one estimate, they could gobble up the entire planet in about three hours flat."[36]

Goo appears as the fear of being engulfed by a monstrous fluid constantly displaced elsewhere, a horror continually alienated, and in so doing, responsibility—or, better yet, paternity—denied: "THEY call it . . ." The nightmare of being swallowed by flows of nanostructures must be distanced and repelled to protect nanoscience's logic of control. According to Ian Gibson, former chairman of the United Kingdom's House of Commons Science and Technology Committee, "We shouldn't be associated with scare stories—science fiction about grey goos and the world being swallowed up."[37] That is, nanotechnology must repel from itself the very fantasy of *global consumption* that it nevertheless in many ways depends on; it must dislocate and separate with horror or derision the very aspiration for a *global ecophagy* (i.e., "shaping the world atom by atom") that is as desired as it is prohibited.

For nanotechnology, then, the gray goo scenario would be a fictitious symptom of its own internal limitations, or insecurities, the symbolic effect of a will to mastery that contains the seeds of its own deconstruction. Which is, of course, why gray goo appears at the site of nanovision's collapse of the real and the imaginary, science and science fiction, the human and the molecular:

> [The abject substance] shows up in order to compensate for the collapse of the border between inside and outside. It is as if the skin, a fragile container, no longer guaranteed the integrity of one's "own and clean self" but, scraped or transparent, invisible or taut, gave way before the dejection of its contents. Urine, blood, sperm, excrement then show up in order to reassure a subject that is lacking its "own and clean self." The abjection of those flows from within . . . enables him to avoid coming face to face with another, spares himself the risk of castration.[38]

The molecular abjection evinced by pundits of nanotechnology in turning away from gray goo—that is, their various attempts to discredit the speculative dangers of self-replicating nanomachines—would seem to work as some kind of coverup operation, plastering over any suggestion of technical insufficiency or failure.

We have seen similar operations elsewhere in nanodiscourse. For example, Binnig and Rohrer have proffered the scanning tunneling microscope as an acquired object that would make up for an originary human lack relative to the nanoworld. Describing their desire to study nanoscale surfaces and "enter a new world," they soon discovered a constitutive shortcoming in human probing prowess: "We realized that an appropriate tool was lacking."[39] It is probably not surprising, then, that in response to this recognition of lack, a *dream* points the way to a compensatory tool. Indeed, a forgotten or repressed dream suddenly comes upon Binnig, tunneling out from the unconscious to show him how to overcome the void: "As a result of that discussion [wherein we realized a lack of proper equipment], and quite out of the blue . . . an old dream of mine stirred at the back of my mind, namely, that of vacuum tunneling. I did not learn until several years later that I had shared this dream with many other scientists, who like myself, were working on tunneling spectroscopy."[40] The "shared dream," the collective unconscious of surface science, unveils a means to supplement the notable absence of an appropriate tool for entering the nanoworld (or effing the ineffable). The assumption of this tool—described in ontogenetic terms as a disciplinary movement "from birth to adolescence"[41]—fills in the hole and lets the subject of probe microscopy enter the symbolic order of "mature" science. In other words, the STM enables nanoscience to grow up by glossing over a primordial inadequacy.

An inadequacy perceived within nanoscience, by nanoscience itself. Which is perhaps why we so frequently encounter expressions of size anxiety in nanodiscourse, a fixation on "very, very small" objects that must seemingly swell to gigantic proportions to gain societal respect. Provocatively, we are told everywhere in nanodiscourse that "size matters" because, of course, nano is all about the question of size: size reduced, deflated, and shrunken.[42] The nanoscientist Thomas Theis says, "When it comes to matter, size matters."[43] But as size would never be a question outside the concerns of matter, or even outside matters of concern, then this is simply saying that size always matters, and vice versa, matter always sizes. So then what's the matter here? Well, nanotechnology might seem to be . . . well . . . a little *small*. But not deficient, mind you! Which is why we are told that, however wee the offerings of nanoscience might seem, things are really *much* larger than they first appear: "There is no doubt that this science of the vanishingly small is the beginning of something very big indeed."[44] "The next big thing is really small."[45] "Nothing 'nano' about it."[46] "Size matters. It's not what you think, but WHAT a difference size makes!"[47] Perhaps it's not what we think. (Are you thinking what I'm thinking?) Though, of course, at the same time, how could it not be what we're all thinking? So just in case it *might be* what we're thinking, nanorhetoric inflates the ostensible size of nanotechnology from the vanishingly small to the turgidly large: an inflation that now seems an *overcompensation* for imaginary shortcomings, aboriginal failings, elemental absences.

Small is big, really. Or, at least, nanotech may be small, but it's got it where it counts: "small science with a big potential."[48] Or, sure, size matters . . . but really, "smaller is better."[49] Or, as the nanoscientist Ted Sargent says, "It's not how small it is, it's how you use it."[50]

These compensatory strategies of nanorhetoric replace lack with exaggerated tropes of superiority, grandiosities of scale, or supplementary probing tools and thereby impose a blind spot at the supposed site of insufficiency. Like these tropes, the social operations of molecular abjection too would serve to produce a defensive coverup. Indeed, Drexler has accused the U.S. National Nanotechnology Initiative of *willful blindness* in its early years of existence for avoiding discussion of molecular manufacturing and its attendant dangers:

In the global race toward advanced nanotechnology, the U.S. NNI leadership has its eyes closed, refusing to see where the race is headed. This creates growing risks of a technological surprise by a strategic adversary, while delaying medical, economic, and environmental benefits. It's time to remove the blinders and move forward with public dialogue and vigorous research, embracing the opportunities identified by Richard Feynman.[51]

Drexler points to the constitutive "blinders" on which a false sense of security in nanotechnology depends: a faith that nano is already well under the control of the "proper" authorities. Certainly, the success of nano in many ways (especially financially, and especially in its cultural infancy) demands these blinders to the imagined threat of self-organizing machines, for acknowledging it would indicate a possible failure of power, a wound already anticipated. Calculated avoidance of this threat—silencing discussion of molecular manufacturing and self-replication as taboo topics within the "legitimate scientific community"—would constitute the kind of structural prohibition on which (patriarchal) authority is itself erected.[52] Whether the NNI's blinders are warranted by "real science" or not, the gag order on goo-talk would indeed be necessary to secure support for the global nano-endeavor. Nanoscience could not allow itself to appear fundamentally dangerous to our human future and still retain public confidence in its ability to control matter. Imposing blinders compensates for what would otherwise appear as premonitory risk, a deflation of security.

Drexler boldly points to fissures in the armor of orthodox nanoscience as represented by NNI policy—for example, the blind spot in U.S. nano legislation concerning research in molecular manufacturing, an absence that draws attention to itself even when, or rather because, everyone has been told *not* to look at it: a palpable "omission."[53] But having done so, Drexler then burns in discursive effigy as the one who is *himself* blinded, accused as the purveyor of "irrational" nanocalyptic fears.[54] "Some of his notions," insists the nanomaterials researcher Chris Ober, "are just plain nutty."[55] Against such accusations, Drexler has suggested that the NNI and the U.S. NanoBusiness Alliance have tried to discredit and silence people like himself who take the dangers (and hopes) of molecular manufacturing seriously: "[We] are indeed being marginalized by those who speak for the nanotech business community."[56] According to Drexler, these fiscal and political stakeholders in the utopian nanofuture "have attempted to narrow nanotechnology to exclude one area of nanoscale technology—the Feynman vision itself."[57] They banish from the "legitimate scientific community" anyone who supports this way of seeing otherwise—this way of seeing that we have been exploring as nanovision.[58] They stigmatize individuals captivated by nanovision precisely to blind out the specular and speculative "Feynman vision in its grand and unsettling entirety."[59] But closing eyes to the insights of nanovision will be an already futile gesture: "Continued attempts to calm public fears by denying the feasibility of molecular manufacturing and nanoreplicators would inevitably fail, thereby placing the entire field calling itself nanotechnology at risk of a destructive backlash."[60] "Those who speak" to accuse Drexler of unreality ("science fiction"), mental failure ("fuzzy-minded" and "nutty"), "impossibility," and "childishness" would speak only to cover up their own fictions, their own inability to guarantee utopian futures or to forestall risks, their own blindness to the many delights of nanovision *as well as* the horrors of goo. In other words, Drexler has recognized himself inhabiting the structural position of the *scapegoat*.

Sacrificing Drexler prevents disintegration of the legitimate scientific community over issues of molecular self-assembly and self-replication; it localizes any larger problems of nanodanger to the fantasies of one man and enables the social field to come together and pretend as if there were no uncertainty about its ability to ultimately control matter. This, as Žižek has written, is the paradox of the scapegoat: "The illusion of the sacrifice is that renunciation of the object will render accessible the intact whole. . . . In other words, what appears as the hindrance to society's full identity with itself is actually its positive condition."[61] The legitimate scientific community *needs* Drexler, *requires* Drexler, if only to abject him, because by forcing him into "the role of the foreign body which introduces in the social organism disintegration and antagonism, the fantasy-image of society *qua* consistent, harmonious whole is rendered possible."[62] Drexler is made excremental; he is found guilty of "unreality" and "science fiction," guilty of "scaring the children" and speaking the "impossible." His guilt then serves to establish the relative innocence of "real nanoscience." His protests against such abuse only help to further distance him from the place of legitimacy: "Guilt is projected onto the scapegoat whose sacrifice allows us to establish social peace by localizing violence. . . . As soon as a man finds himself occupying the place of the sacred victim, his very being is stigmatized and the more he proclaims his innocence, the more he is guilty—since his guilt resides in his very resistance to the assumption of 'guilt,' i.e., the symbolic mandate of the victim conferred on him by the community."[63] Drexler sees himself victimized, but there is nothing he can do. By excreting Drexler from "real nanotechnology," the orthodoxy maintains the illusion that any resistances to nanotechnology, any obstacles preventing the utopian nanofuture from arriving, any hindrances to the closure of the nanocommunity in harmonious accord, any hint of nanotechnology's dangers to itself—any sense at all of nanotechnology's potential for catastrophic autocastration—have been exterminated. Nanotechnology therefore demands this sacrifice of its own father.

Scapegoating Drexler enables "real nanotechnology" to pose as if its very real internal resistances did not exist. "Real nanotechnology" would then achieve a sense of its own rigor and consistency by this very sacrifice, or displacement, of goo-talk. It could then shift risk outside of itself by maintaining that it is the gray goo scenario alone that represents a lack of rigorous thinking. To maintain that thinking "gray goo," indeed, is thinking "soft." "Fuzzy-minded" thinking. "Science fiction" thinking. "Sloppy" thinking. For everywhere in support of "legitimate" nanodiscourse, this kind of wrong-headed imagination about gray goo is itself described in terms of swamps or viral floods. The fluidity of the gray goo problem would seem to infect our very ability to think about it, becoming itself a kind of discursive goo, a spreading "quagmire"—"This is the quagmire we're in"[64]—of misinformed panic. The nanoblogger Howard Lovy describes "ignorance self-replicating" as an effect of hyperreal nanodiscourse, "mistaking the show for the truth" and finding real danger where there is only science fiction: "Meanwhile, the viral spread of nanotech misinformation continues."[65] In 2003, U.S. Representative Sherwood Boehlert similarly said before a congressional hearing on nanotech consequences: "As many people here know, the most extravagant fear about nanotechnology is that it will yield nanobots that will turn the world into 'gray goo.' That's not a fear I share, but I do worry that the debate about nanotechnology could turn into 'gray goo'—with its own deleterious consequences."[66] Self-replicating goo comes to characterize the discourse on self-replicating goo, spewing forth from a disattachment or breakage of reality, a viral infection spreading from a gap in rational thought. For instance, William Tolles, a nanotech consultant and retired Naval Research Laboratory chemist, has said:

> "Visionaries" who have never performed experiments have nevertheless constructed [nanotech] scenarios raising highly questionable possibilities. Due to a lack of contact with reality, they envision a world in which ideal "machines" assemble atomically perfect systems having surprisingly "smart" capabilities. These systems are ostensibly not only capable

of reproducing themselves, but are intelligent, and may be constructed to cause harm to the environment or living species.[67]

Such visions of nanotechnological harm, for Tolles, are explainable only by some constitutive *lack*—a lack of "experiments," a "lack of contact with reality"—because they are "irrational, generated from hypotheses far removed from experiments and the laws of nature as we understand them today" (218). These ideas are "nearly impossible based on the laws of physics, thermodynamics, or other laws of nature, as we understand them" (222–23). These visions—"figments of imagination" (222)—would represent a form of *psychosis*, a total severance or "removal" from any legitimate sense of reality and an incomprehensible rejection of established "laws of nature" ("natural" laws reinforced by *social* consensus: "as we understand them today"). According to Tolles, nanovisionaries have "constructed scenarios raising highly questionable possibilities" and "irrational fictional predictions" that are simply "bizarre" (222, 218, 223). It seems that the "legitimate scientific community" and other self-professed spokesmen of "law" perceive in gray goo a condition of absolute scientific deficiency—a lack or foreclosure of experiments and laws—and an irrational refusal in those who see gray goo as "real science" to accept their lack of (scientific) authority to speak (for) reality.[68]

Likewise, the nanoscientist and Nobel laureate Sir Harold Kroto, responding to some widely publicized statements made by Prince Charles in 2003 about gray goo, asserted that Charles, too, has been a victim of nanovisionary psychosis: "Someone's had this ridiculous idea about nanoscale robots that can replicate themselves, and it's so far-fetched as to be utterly preposterous. . . . It shows a complete disconnection from reality. He [Charles] should take a degree in chemistry, or at least talk to someone who understands it, rather than reading silly books."[69] To entertain the possibility of gray goo is to be disconnected, severed, preposterous (inverting natural order, putting the post before the pre), and silly, while on the other hand, to deny the possibility of gray goo is to be connected, whole, naturally ordered, and serious. To deny goo is to have some *thing* (for example, a degree in chemistry), to hold in one's grasp some emblem, some signifier suggesting that there is no hole in one's field of knowledge, no hole in one's perception of reality. To deny goo, in other words, would seem to show the world that this "someone who understands" (this "one presumed to know") has what Charles lacks. This "someone" has a grasp on that very reality from which Charles has been severed, and therefore the prince should defer to the visible authority of the credentialed nanoscientist. Refusal would be the very essence of madness.

Turning a blind eye to any speculative future that bodes quagmires of goo is therefore to act as if nanotechnology were in complete possession of the "technoscientific phallus": the centralizing principle of control, coherence, and (mathematical) predictability within structures of scientific thought, revealed so often in the telling descriptor of "hard science."[70] If nanotechnology is "hard science," then it is safe: it will not succumb to goo or to soft, science fiction thinking. Indeed, as a *Boston Globe* editorial suggested in 2002, the very question "How safe is nanotechnology?" is answered, and our concerns assuaged, by the fact that trustworthy scientists and businessmen are taking nanoscience and its possible consequences very seriously: "Once it seemed like science fiction. . . . Today nanotechnology is hard science."[71]

Such attempts to represent nanotechnology within the symbolic economy of gendered humanism—specifically, the restrictive two-sex model that has characterized Western thought for the past several centuries[72]—would seem to constitute a tactical response to the injurious threat of the gray goo problem: injurious on the symbolic level to the "laws of nature" that would forbid the existence of such fluidic chaos; injurious on the imaginary level to the social field, where gray goo as a science (fiction) of exploratory engineering, even if manifestly "impossible" or "absurd," nevertheless jeopardizes nanoscience's prosperity; and injurious even on the level of the real, where our entire planet might risk annihilation by the devouring spread of hungry nanomachines. Recasting nanotechnology within a familiar cultural register of male authority and technoscientific stability would, for some, perform an act of inoculation against the affront to traditional humanist thought instantiated by fictive gray goo. To protect its future, then, nano begins to pose as a technoscience that is both *already real* and *always hard*. This posturing is constantly reinforced at the level of popular depictions of nano—not limited to nanowritings from the "legitimate scientific community," but also in fictional films, pop-science books, and television commercials.

What we see here is a particularly gendered form of what Mark Hansen has called *technesis*: the systematic discursivizing of technology that tames and profoundly *humanizes* its radical alterity, diminishing the material impact of its true inhuman otherness.[73] As nanotechnology is put into discourse, it is assimilated to preexisting human gender categories with all their cultural associations, extinguishing those specular and speculative launches of nanovision beyond the limits of man and humanism; this protects human thought from imperilment in the interface with otherwise inhuman materialities of the nanoworld—for example, it wards against "molecular erotics"—as well as the posthuman epistemologies of the nanofuture—for example, it parses "science (fiction)" into mere "science" and "fiction."

I take psychoanalytic theory in its intersection with structural linguistics as a useful inventory of sociosemiotic processes that generate the humanist subject, an accounting of the production of "the human" and its impoverished operations of sexual difference in modernity, which helps us to diagnose the recalcitrant modes of binary thinking that encrust the technoscapes of nanoscience. For as we see, there is a humanism of nanovision that opposes its own posthumanizing tendencies, a desire for molarity that opposes its own molecularities, a logic of control that opposes its own interfacial ethics. By attending to the signifying practices of our culture that construct humanness according to obligatory interpellations of gender and sexuality, or what Judith Butler describes as "the matrix of gender relations . . . that orchestrates, delimits, and sustains that which qualifies as 'the human'"—the same signifying practices that regularly impose and enforce anthropomorphic gender roles even in domains of the otherwise asexual, the abiological, the inhuman, or the self-replicating—we can observe how technocultural negotiations of various nanofantasies try to shore up the leakages in humanism caused by the emergence of nano itself.[74] And if, in the end, nanovision ultimately undermines humanism and its two sexes, this is only by virtue of its working through the multitude of conceptual crises it animates, including the controversies and contradictions of goo.

THE POWERS OF MATTER

The difference between matter in control and matter out of control—that is, the purported distinction between safe, "hard science" nanotech and dangerous, "gray goo" nanotech—regularly appears in contemporary nanodiscourse as a functional differentiation between insemination and its imagined crisis, that is to say, dissemination. Which is nothing new. Across the history of humanist metaphysics, raw matter itself has often been aligned with the feminine and fecundity—as Butler writes, "The classical association of femininity with materiality can be traced to a set of etymologies which link matter with *mater* and *matrix* (or the womb) and, hence, with a problematic of reproduction"—while the forces coordinating matter and enabling its faithful reproduction (form, for example) have been aligned with masculinity.[75] If the presumed passivity or receptivity of matter became historically thinkable only in relation to idealized active forces of form, order, and mechanics, this seemingly indicates an anthropic construction of the solid in primacy to the liquid, or what Luce Irigaray diagnoses as humanism's symbolic "teleology of reabsorption of fluid in a solidified form."[76] The compression and canalizing of matter in its essential movement into stasis, solidity, and permanence—the construction of its proper *ontology*—would thus describe the symbolic economy of humanist metaphysics throughout its overdrawn history. Indeed, the phallic signifier itself symbolizes this ontological containment of matter in motion, material flux. Lacan writes:

> One could say that this signifier is chosen as the most salient of what can be grasped in sexual intercourse [*copulation*] as real, as well as the most symbolic, in the literal (typographical) sense of the term, since it is equivalent in intercourse to the (logical) copula. One could also say that, by virtue of its turgidity, it is the image of the vital flow as it is transmitted in generation.[77]

The anthropic phallus stands for insemination, the emblem of the coordinated symbolic field and the closed circuit where matter and information are not lost or dispersed but always return to their assigned place; or in Lacan's famous formulation of this itinerary: "A letter always arrives at its destination."[78] The phallus is that which would determine the vectoral properties of matter; it would ensure the constraint of matter in being the medium or principle of its distribution. Which is why matter has long been pressed to service those autogenic fantasies of the humanist symbolic regime that take materiality as a passive substrate, a receptacle to be informed by some sovereign force funneling more essential, or seminal, properties.

If insemination has thus been imagined immemorially as the condition of matter in control—the condition of ("natural") law—then any violation of its proper ordering would be criminal, unthinkable, or completely "irrational." For matter to revolt its natural conduction and blockage would be a literally "preposterous" situation for binary thought, or indeed, the deconstructive disorder of monstrosity, as Zakiya Hanafi has written:

> If matter moves of its own accord, it presents a threat of breakdown, of collapsed boundaries, not only in the realm of natural forces . . . but especially as a figure for the stability of the social order. . . . If matter begins to rule spirit, if it broke free and began to exhibit autonomous capacities

. . . the same horror would be unleashed that is created when a body rules soul, or when the resistance of female matter defies the formative virtue of male semen in the womb, or when spirit is called into inappropriately formed matter. What would happen is that monsters would be generated.[79]

The very idea of matter out of control, the self-organization of material flux—matter that disseminates, in all its senses—appears as uncondensed horror. For matter that moves and reproduces itself, deterritorialized from form, essence, or logos, disintegrates the humanist symbolic structure when the autocratic signifier of insemination—as the very emblem of logical *copula*, the anthropocentric sign of being conceived in the vectoral image of vital flow—is monstrously engulfed by flow as such.

For nanotechnology, monstrosity materializes as the autocatalyzing molecular machine, a fictive violation of the "laws of nature," an illicit revolt of matter itself, a literal *devouring* of technoscientific sovereignty from below.[80] In other words, all this "science fiction about grey goos and the world being swallowed up," as some nanoboosters put it—all those dark spaces of "mom's basement" and childhood bedrooms, ridiculed by the legitimate scientific community as incubators of irrational goo fantasies—would represent molecular abjection, a revolution of the mothermass, or what Barbara Creed has termed the "monstrous-feminine."[81] The nightmare image of the devouring flood of nanotech goo, sucking vital essences into itself, giving birth to self-replicating machines even as it brings about the death of the biosphere, this revolting matter would mark *fundamentally*—as desire and eschatology— the *end of man*.

We see this interplay of longing and revulsion in Klaus Theweleit's account of the protofascist "soldier male"—his exhaustive psychoanalysis of German Freikorps troops between the world wars—and the imaginary function of floods, fluids, and revolutionary masses (indeed, goo) as signs of matter out of control that threaten the autonomous masculine subject with complete dissolution.[82] To combat the loss of mastery perceived in women's bodies and in mass revolt, the soldier male responds by simultaneously taking on a sense of hardened body armor and turning other bodies into pulp, directing the war machine against fluidic masses to prove they cannot penetrate, cannot contaminate, cannot disintegrate. In becoming an armored body bristling with explosive and piercing weapons, reasserting control over those inner flows of goo within himself which he must displace to the outside, the soldier adopts a mechanistic mentality, as if to say, "I am a machine." The psychodynamics of Theweleit's mechanized soldier protecting himself against chaotic matter provide an important insight on the discursive practices of nanotechnology.

In the dreamscapes of nano, the threat of goo is equally the threat of the machine, both the nanomachine yet to come and the nanomachine already inside (Whitesides: "The once and future nanomachine"), the "natural" molecular machines that already make up our cells (Montemagno: "Within all living organisms exist the original nanomachines"), and the small nanoworld contained by the subject of nanovision as haptic space, domestic interiority (Brauman: "There is, indeed, room at the bottom, and we are beginning to move in"). Protection against the machinic on all sides, inward and outward, past and future, would then involve assuming technical control—taking up the armor of "hard science," performing technoscientific acts of insemination. Indeed, some nanoscientists have been trying to harness the motor proteins of mammalian sperm and funnel them into engineered devices, using "sperm power" as a tool to control their molecular machines: "We're taking what sperm have already figured out how to do and using it for a nanotechnology application."[83] Such displays of technical prowess firm up the armor of hard science, channeling the flows of matter through proper experimental laboratories while displacing rogue forms of nanotechnology outside as soft, irrational, fictitious, and other.

Which is why gray goo figures as abject matter nearly everywhere in the narrative texts of nanoculture relative to a technoscientific humanism that strives to combat its dispersion. Aiming for containment, the dominant perspectives represented in or by these texts comprehend self-replicating molecular entities through the register of the human and its exclusive forms of sexual difference, and the contradictions and aporias that inevitably thereby arise are often foreclosed with mere obstinacy, if not reactionary violence. In Ian McDonald's novel *Evolution's Shore* (1995), a goo of extraterrestrial origin begins to devour Africa, transforming the land into a bizarre alien world of buckyball structures, self-replicating "fullerene machines" and hybrid biomechanical creatures.[84] Named the Chaga, this nanotech goo is a spreading, fluid infection: "It invades the land, draws strength from it, kills what it finds and duplicates only itself. . . . It is growing, it is spreading" (16). It thwarts efforts to confine it while also defying heterosexual reproductive imperatives. As one scientist in the novel describes it: "The Chaga may just be dumb, fecund life, with no more intelligence than . . . a condomful of sperm" (81). The Chaga appears to be fecundity *without* insemination—it embodies *dissemination as such*—representing the loss of voice ("dumb") and of "intelligent" or rational meaning, that is, the loss of *presence* itself in the onanistic scattering and joyously wasteful dispersal of fluidity and flow.

As the very stuff of molecular abjection, Chaga emerges at the site where boundaries between self and external environment dissolve, a tremulous gap between symbolic and real: "Not so far on the map but on the very frontier of the imagination, the Chaga" (18). When consumed by Chaga, the human body is no longer itself but instead interfaces in "symbiosis—that is the way of the Chaga, to join with things not of itself and draw them into it" (61). The Chaga devours human beings only to regurgitate them as hybrid entities now intimately fused with itself, human no more but bleeding together with Chaga: an abject molecular erotics, where Chaga and human "lovers" know each other "as intimately as any one thing may know another" (61). The bleeding together seems both desired and repellent: "Wonder and horror. Beauty and terror. Paradise, I suppose, but at the same time . . . an insidious hell. Monstrous, wonderful" (55). This molecular abjection, or erotics, bursts forth precisely when imaginary skin fails and the subject becomes continuous and undifferentiated from its surroundings, identical to the postmortem or the prenatal. As one man who has been swallowed by the Chaga expresses it:

> When you wake from death, it is all the terrible things about being born again. You are forced from a warm, comfortable womb of flesh into a world that you cannot understand. I woke in darkness, kicking at the bubble of soft skin into which I was curled. It unfolded around me like a flower. . . . In one of these I came back to life, halfway up the face of a huge reef of Chaga growing out of the mothermass. (58–59)

The Chaga consumes, digests, and regurgitates new hybrid life, now undifferentiated from the "mothermass." The Chaga is a vast and spreading "warm, comfortable womb," eating the world and giving birth to something different. One nanoscientist in the novel clarifies: "Essentially, the Chaga is one mother of a buckyball jungle" (79). While some characters intuit the Chaga's transformative power as a kind of "essentially" maternal "womb magic" (211), it is also evident that this alien force explodes the human taxonomic grid as such, demanding new modes of thought: "It's the end of everything it means to be human. . . . Or a gate into new ways of being human" (213). But as surely as it instantiates an "end" or a "gate," a goal or a tunnel, a closure or a surge, this singular peril (or promise) to human being also incites loathing and global demands to quarantine it, whether in governmental and scientific efforts to police its physical boundaries or in the concomitant discursive efforts to police its gender, efforts that arguably amount to the same thing.

The anthropocentric comprehension of nanotechnology as a surging and devouring womb recurs similarly in Wil McCarthy's *Bloom*, wherein a nano-goo known as the Mycosystem has taken over our inner solar system, forcing the human survivors of a doomed Earth to flee to the moons of Jupiter for safety. Establishing a lunar colony called the "Immunity," mankind decides to make efforts to reclaim the homeworld, organizing a mission back into the Mycosystem. The novel's narrator, John Strasheim, is selected to accompany this mission as a reporter. He prepares for departure by visiting his mother while contemplating his upcoming voyage into the Mycosystem: "Could I do that, leave Ganymede behind, leave the whole Immunity behind to dive back into the warmth of the inner solar system?" (14). In this early chapter, appropriately entitled "Wombs," visiting one's mother and flying into the Mycosystem are both figured as returns to the warmth of the womb—and simultaneously as abandoning the masculine "immunity" or sterility indicated by the sign of "Ganymede," the homosocial lunar space of "Ganymede behind."

This punning sexual tropology is played up throughout the mission into the Mycosystem, where the ship must be protected from the hostile machinic environment by a thin sheath of nanocomposite that makes it immune to the infectious and hungry goo. This strategy is described as "a sort of Trojan horse approach" (182)—and the prophylactic association is quite evident. Moreover, the mission's purpose is to deliver "detector probes" to the goo-encased planets of the inner solar system, including "Mother Earth" (255), attached like a mammalian egg to the monstrous uterus of the Mycosystem. It soon emerges that "the 'detectors' to be seeded across sterile portions of the inner planets are in fact ladderdown explosive devices intended to devastate the surface of the worlds" (59), thereby cleansing the solar system of nanotech. In other words, the little ship will spill its "seeds" into the Mycosystem and, through this act of insemination, reclaim man's territory. Strasheim imagines their entry: "a translucent Mycosphere, enormous, its upper boundary rippling with our penetration, its lower one far below, the haze of it obscuring the yellow smiley face of Sol at the center. And ourselves? A single pixel plunging down through the ripples" (177). That this cosmic mass is ultimately unaffected by the little pixel's plunging is all the less surprising when it becomes clear that the Mycosystem is quite beyond these merely discursive efforts to humanize it, or domesticate it. But the human crew of the mission nonetheless imagines its role as penetrator and inseminator, and the Mycosystem's role as receptive womb to be seeded and cleansed of goo, as the acutely *human* solution (psychological, if nothing else) to the revolt of nanotechnology.

Likewise, in Crichton's *Prey*, not only do wild nanobots pose a gooey threat in themselves, but the very field of nanoscience appears to constitute an immediate affront to the modern nuclear family and its paternal center. Jack Forman is an unemployed computer scientist who must stay at home and raise his children—he is a "house husband"—while his wife Julia, a successful nanoscientist, supports the family. Devoted to her work, Julia spends too much time away from home, and Jack comes to believe she must be having an affair; he even sees her riding with a strange man in her car. This "man," however, is really an intelligent swarm of gray goo nanites, created by Julia's drive for professional and financial success. This "nanoswarm" takes up residence inside her body, effectively displacing Jack's spousal claim for access to her inner parts. No mere sexual rival, though, the nanogoo is a kind of monstrous progeny, a swarming fluid that extrudes from Julia's body and attacks the patriarchal family, including Jack and Julia's children. Julia is evidently an "unfit mother," sacrificing her (human) family for nanopower. As in other Crichton novels that animate white male paranoia in the postmodern world—such as *Rising Sun* (1992), which posits the secret Japanese takeover of corporate America, and *Disclosure* (1993), which displaces the problem of sexual harassment onto a female employer and her victimized male employee—*Prey* stages the unmanning of a father and the collapse of a nuclear family as immediate consequences of entrenched social orders reversing themselves.

But it turns out that woman's nanogoo, menacing the whole world, can be resisted only by this same unmanned father, Jack Forman, who struggles in the name of the father, or, as his surname would suggest, "for man." At the climax of the novel, rather than entertain the nanoswarms' visionary proposals for a "new synergy with human beings," a new age of human-machine symbiosis, Jack decides to destroy them all.[85] He describes how he forces the swarms and their human hosts into submission: "They screamed as the water touched them. They were writhing and beginning to shrink, to shrivel right before my eyes. Julia's face was contorted. She stared at me with pure hatred. But already she was starting to dissolve. She fell to her knees, and then onto her back. The others were all rolling on the floor, screaming in pain" (355). Moments later, the swarm-human symbionts are entirely melted down by a thermite blast, their horrors now exterminated: "You could almost believe it never happened" (357).

Popular fictions such as McDonald's, McCarthy's, and Crichton's thus thematize and appraise various gendered representational strategies for neutralizing the danger of nanotechnology's resistance to control and its sheer alienness to the matrix of humanism—the same strategies, it seems, as deployed by "legitimate" scientific discourse. I will trace these strategies of domestication in greater detail across three generically distinct texts: the blockbuster film *Terminator 3: Rise of the Machines* (2003); the pop-science book *Nanotechnology and Homeland Security: New Weapons for New Wars* (2004); and finally a commercial for General Electric washers and appliances, "Beauty and Brains" (2003), which became momentarily famous as the first advertisement broadcast on American television to showcase the word "nanotechnology."

"DESIRE IS IRRELEVANT"

Although nanotechnology does not appear in the *Terminator* saga until the third installment, *Terminator 3: Rise of the Machines* (dir. Jonathan Mostow, 2003), the earlier films predetermine nano's relation to the anthropic regime. In the original *Terminator* (dir. James Cameron, 1984), a killer cyborg (played by Arnold Schwarzenegger) travels back in time to the Los Angeles of 1984 to exterminate Sarah Connor, the mother-to-be of John Connor. In 2029, John Connor will lead the human resistance against autonomous machines that "rise from the ashes" following a nuclear cleansing of the world by Sky-Net, a self-aware computer defense system who "began to see all humans as enemies." By sending the Terminator to kill the mother of its primary enemy, SkyNet hopes to alter the future and destroy all traces of human resistance. The Terminator, as the killer of mothers, babies, and fathers, is not only resolutely antihuman but also antireproductive and antipaternal, violently refusing heterosexual imperatives.[86] The cyborg embodies sexual ambiguity and the destruction of patriarchal order, targeting the police and the military and all other icons of "law."

"There's a storm coming," the film tells us. A metaphor for dissemination and techno-chaos, the storm will begin in the future when SkyNet leads autonomous machines in an all-out war against humans. Surprisingly, this storm can be opposed only by a family drama involving Sarah Connor successfully giving birth to John Connor (whose initials, JC, are no coincidence), a drama that owes its success to the efforts of Kyle Reese, John Connor's friend in the future who travels back in time to protect Sarah from the Terminator.[87] Ultimately, this first *Terminator* film leads its audience to believe that machinic rebellion will be stamped out by John Connor, whose birth is ensured by what proves to be a fantastic act of *insemination*.

"Do I look like the mother of the future?" Sarah asks. The mother of the future gives birth to the future through a terminal looping of information: knowledge of the inevitable birth of her son ("At least now I know what to name him") and genetic information carried by Kyle, the father of her son, sent through time by the son to become his own father. Kyle says to her:

> John Connor gave me a picture of you once. I didn't know why at the time. It was very old, torn, faded. You were young like you are now. You seemed just a little sad. I used to always wonder what you were thinking at that moment. I memorized every line, every curve. I came across time for you, Sarah. I love you. I always have.

Reese, in love with the photographic image of an already-pregnant Sarah, announces: "I came across time for you." This startling double entendre initiates the very love scene that leads to Sarah's pregnancy. She becomes "the mother of the future" only because Kyle "came across time," inserting both semen and futuristic knowledge into her body. The future impregnates the past to give birth to itself, thereby saving the world for mankind: the coming storm arrested by a tightly contained human orgasm.

This fantasy of an impossible phallus that can extend itself "across time" to inseminate a woman (even one's own mother!) in the past bodies forth the Oedipal structure of the film.[88] *The Terminator* suggests that anthropic insemination, able to channel matter and information into the receptive bodies of terminal women, ultimately outcompetes the cyborg storm. This subtext is more than reaffirmed by the second film in the series, *Terminator 2: Judgment Day* (dir. James Cameron, 1991), which features a "liquid metal" metamorphic T-1000 as the new machine villain, whose fluid powers are ultimately defeated by the hardened original Terminator now working in service of humanity (even playing the role of substitute father for John Connor).[89]

Terminator 3: Rise of the Machines once more figures technological apocalypse as an affront to patriarchy, but where the previous films portrayed our machinic enemies as sexually ambiguous or fluidic "males," *T3* depicts their threat as thoroughly feminine.[90] The newest Terminator model is the T-X, metamorphic like the T-1000 but whose preferred form is that of a blonde Caucasian woman (figure 17). Dubbed "the Terminatrix," this cyborg is also a source of nanotechnology, which bubbles forth from her body to cause further machinic chaos. The Terminatrix, it seems, is a Terminator with a

17. The Terminatrix, played by Kristanna Loken. (*Terminator 3: Rise of the Machines*, Warner, 2003)

"matrix," or womb. While the technology of the original Terminator embodied a recognizable masculine violence in its relentless pursuit of Sarah Connor—suggested by repeated insistence that the Terminator is a "fucker" and a "motherfucker"—the Terminatrix embodies a feminine violence describable only by an inadequate condemnation of "proper" womanly behavior: "You bitch!"

The T-X's threat of "bitchy" violence is nevertheless infinitely more lethal than the "motherfucking" violence of the first Terminator or the T-1000. The T-X is an "anti-Terminator Terminator": she destroys both humans and other Terminators. But she especially targets human males in positions of authority, such as policemen, or the heroine Kate Brewster's fiancé (who she believes "will come to rescue [her]"—but whose pathetic demise is suggested only by a splatter of blood across a photo of the doomed heterosexual couple), or Robert Brewster, Kate's father, a military scientist and director of cybernetic weapons research for the U.S. Air Force. Of course, the T-X's main targets are John Connor, the savior of the future, and his heroic father-substitute, the Arnold Terminator (as John Connor says to the T-101, "You're the closest thing to a father I've ever had").

The Terminatrix is armed with a variety of weapons, the most significant of which are "nanotechnological transjectors." Clarifying what the T-X's nanotechnological transjectors enable it to do, the T-101 says: "It controls other machines." As a viral, insidious co-optation of other technologies, the concept of "nanotechnological transjection" goes right to the heart of the gray goo problem, which fantasizes the exponential and unstoppable conversion of innocent and "good" matter into rebellious nanomachines, the dedifferentiation of decent individuals by the riotous mass. The Terminatrix spreads techno-rebellion by injecting her nanofluids into otherwise law-abiding machines; for example, we see the nanotransjectors coursing across the silicon motherboard of a police car, subverting it to her desire.

The film explicitly analogizes nanotechnological transjection with the techniques of feminine seduction. The machinic femme fatale "transjecting" other machines to do her bidding here recalls an entire iconographic history of the cinematic dangerous seductress, figured first in the T-X's deceptively lethal nudity when she arrives in the present, and later in her severe choice of outfit. The Terminatrix dresses in a suit of tight red leather, reminiscent of the fetishistic garb worn by other Terminators in the film series, but with a significantly normalizing difference. The ostensibly male Terminators challenge heteronormative sexuality in their sartorial choices. For example, the T-101 prefers the guise of the "leatherman" (figure 18), evoking a latent queerness even in his earliest appearance despite—or precisely because of—his turgid physique.[91] This latent queerness becomes totally manifest in *T3* when the Terminator appropriates his leather uniform from a "bitchy" gay stripper who tells the cyborg, "Patience, honey! . . . Bitch, wait your turn!" and "Talk to the hand!" After stripping the stripper of his leathers, the Terminator emerges from the strip club wearing this outfit, in perfect sync with the chorus of the Village People's hyper-camp anthem "Macho Man" (1978). The cyborg later adopts "Talk to the hand!" as its own catchphrase (figure 19). Similarly, we might note the queer fetishism of the T-1000, who prefers a police costume (figure 20). (The two male Terminators could almost front a Village People tribute band by themselves!) But the Terminatrix's outfit diverges from the queer male associations of other Terminator uniforms by conforming to the more heterosexual context of the female "dominatrix." As dominatrix, the Terminatrix accommodates existing fantasy regimes of female sexual authority over heterosexual males. She may be threatening, but she does not challenge entrenched sexual categories. Rather, she affirms them. The nature of her sexual "domination" is highlighted when, only moments after donning her outfit and speeding away in a stolen car, the Terminatrix is pulled over by a policeman.

Noticing a billboard that features a voluptuous woman, the Terminatrix causes her own breasts to inflate, as if she intends to seduce the police officer. We have seen similar scenes in innumerable Hollywood movies, and the film relies on our expectations that the T-X will simply flirt and thereby escape the otherwise restraining authority of the police force. She must seduce in order to continue her search-and-destroy mission. Her breasts grow, and we see the policeman's face brighten instantly: "Lady, do you know how fast

18. The Terminator as leatherman, played by Arnold Schwarzenegger. (*The Terminator*, Orion, 1984)

19. "Talk to the hand." The Terminator gets a new outfit—and a catchphrase. (*Terminator 3: Rise of the Machines*, Warner, 2003)

20. T-1000 in police uniform, played by Robert Patrick. (*Terminator 2: Judgment Day*, Tri-Star, 1991) you were going?" Focusing her gaze on his crotch, she says, "I like your gun." So far, this short scene plays to cliché, working up front to recall gender stereotypes: the male authority figure, the voluptuous female, and the limited power the seductress has in drawing attention to phallic power—"I like your gun"—finding her own strength by reflecting the male's desire back to him (what Joan Riviere famously called "the masquerade").[92]

But the scene suddenly subverts these stereotypes by cutting to the T-X continuing on her way, presumably having summarily slaughtered the cop. In the policeman's off-screen death, we comprehend that this dominatrix's power is such that she can take what she wants; she does not need male permission to touch his gun. The scene's seduction dynamic is clearly irrelevant to the plot (the mammary inflation especially seems pointless for a machine who packs such an awesome arsenal of assault weapons! That she does not simply destroy the cop outright only proves the primarily symbolic purpose of the scene). Yet it serves at one level to situate the dangers of cyborg technology within the realm of feminine seduction, the risks of male hardness softening and giving way in the presence of woman. At the same time, this brief scene instantiates the antinomy of two different symbolic regimes: a humanist symbolic constituted by a binary logic and its (anatomical) signifiers of sexual difference, and a posthuman symbolic constituted by fluid processes and the ability to take on or take off genders as desired, to produce or reproduce sexual differences as camouflage or game. The T-X, like the other Terminators, takes on gender simply to exploit the limitations and vulnerabilities of the humanist symbolic system that categorizes bodies and locates desire along a single axis of differentiation, fatally blinding itself to other possibilities. So while the film's visual reinscription of heteronormative gender categories seemingly functions to humanize machinic alterity, it simultaneously suggests that this signifying practice is itself a mode of self-delusion, an inherent blindness of phallogocentric humanism that heralds its own downfall.

Nanotechnology's part in this becomes clear when, in the following scene, we witness the nanotechnological transjectors put to work, squirted into police cars, where they spread and conquer. The T-X's syringe-drilling device for delivering her transjectors, while perhaps an echo of the "phallic woman," is nevertheless no signifier of insemination, for the transjectors spill all over and run riot, diffusing from car to car like a plague, and through this miasma of corruption the T-X now commands a fleet of literally "unmanned" authority vehicles. In both instances of the T-X's subversion of the police—the first, a softening, subduing temptation of flesh, and the second, a disseminating, viral infection of metal—the monstrous-feminine as the temporary guise of the machinic realm demolishes whatever aura of security clings to policemen, their guns, and their cars.

The relation of nanotechnological transjectors to the police here enfolds the problem of gray goo in a very specific way, for in theoretical discussions of nanotechnology, the "solution" to gray goo proposed by several thinkers has been "blue goo"—that is, nanobot police. First imagined by the nano-theorist Alan Lovejoy in 1989, these "bugs in blue," self-replicating authority robots that would keep other nanobots from misbehaving, would seem a logical defense against autonomous gray goo; our security, writes Lovejoy, would depend on having "the world *infested* with repair and defense nano-agents, at high density, ready to spring to action at the first sign of inimical activity [from gray goo]."[93] Other exploratory engineers, including Freitas, have advanced designs for effective blue goo that would involve greater degrees of command over the police themselves:

> Once the malfeasance [of replicating gray goo] has been detected an over-whelming mass of blue goo "enforcers" can be rapidly imported from a central distribution facility. Fresh blue goo may be withdrawn from massive floor-to-ceiling stockpiles in warehouses, wakened and programmed, transported to the site, then slathered atop the infected area as thick as butter on toast, quickly smothering the slowly replicating grays. . . . Hence, to a crude approximation, a single aggressively armed police nanorobot can disable $^-10^9$ unarmed self-reproducing gray-goobots within a single gray goo replication cycle. . . . Obviously the hardware and software comprising the policebots must be *extremely* fail safe and robust, but at least this is subject to our design control.[94]

Blue goo, we must believe, would be strong, robust, immune to corruption or failure, resistant to the influences of gray goo, and fully under "our" control. But as Lovejoy warns, blue goo would have the same inevitable limitations of the human policing agents on which it has been modeled: "Of course, the problem then becomes the reliability and security of the nanopolice. Can they be trusted? Subverted? Could a traitorous strain be introduced that would out-replicate the bugs in blue, displacing them from the target area, and then striking?"[95] That blue goo could be subverted or prevented from carrying out the law when faced with the lawless grays, who might "out-replicate" the police, spread faster, dissolve boundaries, seduce or transduce all other matter, even absorb other good nanobots into its gooey morass, suggests the same threats to authority at the nanolevel as those ex-

21. The severance of the Terminator. (*Terminator 3: Rise of the Machines*, Warner, 2003)

perienced at the psychosocial level. *T3* implies that to place our faith in "blue goo"—like placing our faith in the instruments of the police, or the government, or the military, or any of these "secure" institutions that are ultimately unmanned by the Terminatrix's nanotechnology—is to misrecognize the threat of machinic seduction. Blue goo would be seduced by gray goo, just as police cars and policemen are seduced by the Terminatrix: the powers of an inhuman agent who, even if she "likes your gun," already contains more transvasive technologies of her own.

Even the stalwart Terminator is corrupted by the Terminatrix: symbolically castrating him—she seizes his crotch only moments before severing his head (figure 21)—she then consumes him with enslaving nanotransjectors. Turned against his will, turned against his programmed mission as the father-substitute and protector of John Connor, the Terminator lurches toward his "son," having been once again commanded by the machinic regime to terminate John Connor (the word "terminate" flashes across the T-101's visual screen, alternating with "abort," suggesting at one level the machine's efforts to abort the command, but at another level referencing the mission of the original Terminator: to kill Sarah Connor and thereby perform a "retroactive abortion" of John Connor).[96] The Terminator can only voice the fate that has befallen the phallic instrument after being corrupted ("The T-X has corrupted my system!") by the forces of the machinic future: "Desire is irrelevant. I am a machine."[97]

Nevertheless, despite the demonstrated power of nano-seduction over lesser technologies, the Terminator regains control over himself, repossessing and violently reasserting his (masculine humanist) desire to defend his symbolic son (which is also perhaps a homoerotic desire, since we are told earlier in the film that the T-101 has been "dicking with [John]"). The Terminator, both father figure and queer male, simultaneously possessing and desiring (to serve and protect) the anthropic phallus (presumed to be possessed by John Connor, who will save the human race from machines), in the end successfully resists femininity — *and thereby resists his own machineness.*

Restored as hardened warrior, no longer mere machine but now a real boy with "relevant" desires, the Terminator takes a most symbolic revenge against the Terminatrix: he drags her kicking and screaming on the ground toward him, locking her in a violent embrace and cramming his "fuel cell" into her mouth as if committing an oral rape (figure 22). "When ruptured, the fuel cells become unstable," we are told, and they explode with nuclear force (figure 23). When the Terminator's fuel cell detonates, it takes out both cyborgs and half a military installation: "You are terminated," growls the Terminator in the instant before blastoff (figure 24). (The line was filmed as "Eat me!" at one point but was reshot before final editing.)[98] This last, desperate act of a seduced, adulterated, "corrupted" male against the being it holds responsible for its unmanning works to demonstrate the resurrection, in other words, the *relevance*, of the technoscientific phallus even in its nuclear fragmentation. As Chris Hables Gray writes: "The cyborg penis is quite the fitting sign for postmodern patriarchy: threatened, inconsistent, yet in the end adaptable and bolstered by technology, and so still functioning."[99]

But unlike earlier installments of the *Terminator* saga, where anthropic power would seem to secure an ultimate defeat of the machines, *T3* concludes on a deeply ambiguous note. Having shown not only the failure of the police against the nanotech-feminine but also the nearly complete destruction of the global military by SkyNet, the film suggests that the machine era of cyberspace and nanotech disempowers (military) men at the supposed site of techno-authority. This is made especially clear in the film's attention to the mortal failure of Robert Brewster, father figure and military scientist, who has, we are told, a debilitating "performance anxiety" when facing off with the disembodied SkyNet system, a quivering sense of inadequacy relative to this becoming-entity who has no center but can infiltrate everywhere in the global electronic network: "By the time SkyNet became self-aware, it had spread into millions of computer servers across the planet. Ordinary computers in office buildings, dorm rooms, everywhere. It was software — in cyberspace. There was no system core. It could not be shut down." This soft technology without center, without core, without logos or phallus, spreads beyond control and takes control. It is the very principle of molecular abjec-

22. Disciplining the Terminatrix. (*Terminator 3: Rise of the Machines*, Warner, 2003)

23. Unstable fuel cell. (*Terminator 3: Rise of the Machines*, Warner, 2003)

24. "You are terminated" (a.k.a. "Eat me!"). (*Terminator 3: Rise of the Machines*, Warner, 2003) tion. A *revolting* technological flow extruded from a human source, SkyNet bursts forth with a certain *uncanny* sense of the already inevitable, a certain sense that molecular abjection was already in the works archaically, primordially, in advance of its dissemination: "The attack began at 6:18 p.m.—just as he [the Terminator] said it would. Judgment Day. The day the human race was nearly destroyed by the weapons they built to protect themselves." The same fear recorded in nanodiscourse—the fear that our "ultimate control of matter" will preordain "matter out of control"—fully suffuses the *Terminator* saga. Yet the films remain reassuring in depicting technology within reliable registers of humanism, and in showing that, even if cyborgs and their nano-seductions may represent a real threat to human life, they can be disciplined: just stick a fuel cell in her mouth.

For even as SkyNet is "spreading," John Connor steps to the fore and, anticipating his destiny as savior of the human race in its battle against the machines, poses as androcentric authority. The hopelessly fragmented military forces ask, "Who's in charge here?" and John Connor quietly responds, "I am." At the end of the apocalyptic film, then, the masculine *ego* or *cogito* (the phallus as *copula*) erects itself against technology—"I am," "I am in charge"—anticipating its future might, omnisciently reassured that men will win eventually, because the fathers—Kyle Reese and the Arnold Terminator—have delivered this knowledge to the son when they "came across time." In the film's final lines, John Connor intones: "All I know is what the Terminator taught me: never stop fighting. And I never will. The battle has just begun." His words gesture to a future of endless warfare, of combat that literally never ends because even at that extradiegetic point in the future when humans defeat the machines, the war travels back in time through the temporal-dislocation device and plays itself out all over again. The "end" of this anticipated future loops back to the beginning of the first *Terminator* film—it is therefore not an end—and the future gives birth to the past, just as the terminal fathers said it would. A totalizing circuit of mechanized warfare, locking men and machines into eternal repetitions of mutual victimization: the last resort of the soldier-male mentality when faced with the spreading chaos of the nanotechnological world.

HOMELAND SECURITY

If one strategy for domesticating nanotech has figured the problem of self-replicating nanosystems as monstrously feminine but ultimately controllable, an alternate approach has insisted on the value of nanotech for strengthening technocracy. For example, where nanorhetoric meets the militarized security state, nano becomes a science of armoring and arming the "soldier of the future," the center of the army and therefore the center of homeland security.[100] The body of the American soldier, for one, is being reassembled even now by nanoresearch undertaken by the U.S. Army's Future Force Warrior program and the MIT Institute for Soldier Nanotechnologies, and, as promoters of these extremely well-funded programs assure us, nano-technology is anything but "soft science."[101]

Consider the propagandistic *Nanotechnology and Homeland Security: New Weapons for New Wars* (2004), by the fraternal team of nano-entrepreneur Daniel Ratner and nanoscientist Mark A. Ratner.[102] This pop-science book about American military nano and its civilian benefits presents a vision of the world in constant danger of collapse, a world filled with uncertainty, terror, and inevitable loss which only nanotechnology can prevent. Our small postmodern world—our "global village" (131)—is fraught, we are reminded, with catastrophe: "We face a wide variety of threats. These include the usual suspects: earthquakes, storms, tornadoes, fires, explosions, and spills of toxic materials and pollutants. But threats to homeland security now include terrorism and terrorist acts, perhaps involving biological, chemical, or nuclear agents (often called weapons of mass destruction)" (xviii). Our homeland security depends not only on containing these external and foreign dangers—terrorism, unnatural pollution, and the inhuman forces of nature itself—but also on buttressing our internal domestic structures and systems: "Homeland security involves more than dealing just with toxins, tornadoes, and terrorism. It also involves providing economic, environmental, and educational security" (xix). Everywhere, inside and out, the small world appears on the verge of constant disintegration, and this fact, we are told, has nowhere been made more evident than in the terrorist attacks of 9/11, which opened American eyes to evidence that the world as such was out of whack and that American power was fatally compromised, popped like a bubble: "An attack the size and scope of 9/11—carried out just as America's economy was entering a recession after its euphoric 'tech bubble' years—resulted in a great awakening to the fact that all was not right in the world and that we were not, even on our home soil, anywhere near so safe and omnipotent as we had thought" (4).

This rupturing of the American sense of omnipotence, euphoria, and internal economy—like rupturing a Terminator's primary fuel cell—renders everything unstable. It forces us to see the world as essentially chaotic, with injury looming everywhere. All modern threats to inner robustness and outer stability—"toxins, tornadoes, and terrorism," the fragmentation of economies and educational systems, hazardous elements both alien and native—entail the possibility of global meltdown. Faced with these numerous options for triggering an unstoppable spread of calamities, the authors ask: "How do we work quickly and cleanly to avoid having matters spiral out of control the way they did when a terrorist's bullet started World War I?" (4). And here the paranoid logic of homeland security reveals itself: without absolute discipline, "matters spiral out of control."

Terrorism, global war, inhibited motion, pollution, and the dangers of nature itself all inhabit this concept of "matters spiraling out of control," we see, for homeland security would oppose all such risks by moving unhindered, "work[ing] quickly and cleanly," warding against dirt and contamination, thereby preventing anarchy, world wars, and all other aspects of material pandemonium. To preclude matters—or, indeed, matter as such, the very site of opposition to one's own proper and clean self—escaping our mastery. In other words, to avert all these risks to our secure sense of selfhood and attacks on "our home soil," homeland security must enact something very much like the "ultimate control of matter."

Which is, of course, why nanotechnology appears as the delightful solution to all risks of "toxins, tornadoes, and terrorism," all risks to domestic soil. Even if the soil itself were to turn on us—for example, in the form of earthquakes, tornadoes, or other coming storms—if matter *as such* were to escape our "safe and omnipotent" grasp, nanotechnology would save us. Because if we achieve such technoscientific mastery, then matters will never again spiral out of control. And we can rest assured that our security is already well in hand, the authors tell us, because the control of matter "is not just the technology of tomorrow—nanotechnology applications are already reality" (xix).

If nanotechnology already offers "immediate applications for defense and security" (7), then it would be absurd to imagine nanotechnology *itself* as a security risk. Predictably, the authors insist that nanotechnology is not "the concept of a molecular assembler. . . . [Nor is it] the concept of 'gray goo'" (14). Both these Drexlerian visions are hopelessly flawed, the text insists, and they represent precisely the kind of irrational fragmentation that homeland security works against: "After a bit of closer scrutiny, though, the cracks in this argument [for the plausibility of assemblers] begin to show. There are a number of compelling reasons why molecular assemblers are either impossible or are at best in our distant future" (15). The argument for assemblers and goo is itself "cracked." It is a fissured, "crack-pot" theory:

While the gray goo scenario and its derivatives are certainly frightening ones . . . they are emphatically not what nanotechnology and nanoscience are about. . . . In short, the gray goo scenario is far-fetched . . . Functions this complex simply don't work at a scale this small, and even if nanobots were the size of bacteria or viruses, we have the human immune system and a host of other technologies for defending ourselves against them. (18)

Shotgunning counterclaims at the plausibility of nanobots and gray goo, the authors insist that the idea of goo is "cracked," "far-fetched," and "doesn't work," but also that even if it does actually turn out to work, it would do so only "in our distant future." So it's a problem we can always displace elsewhere: we can hide it under the bed as an irrational dream, or in the future as irrelevant to our current concerns. And should it ever become a real problem, at that time we will be protected by our own human bodies and "other" technologies. Goo would seem to be both impossible and yet possible, both banished to an inaccessible subjunctivity ("even if nanobots were . . .") and yet also dismissed in the present tense as a manageable opponent ("we have . . . a host of other technologies for defending ourselves"). Regardless, nanotechnology is not that! Nanotechnology is not goo! Because goo is science fiction, goo is out of control, goo is *soft*, while nanotechnology is "already reality," it is in control, and, as we quickly learn, it is *hard*.

The nanotechnology applications described by the Ratners rely deliriously on concepts of hardness, cohesion, impenetrability, and the erectile function of nanomaterials. For example, we are told that the nano-enhanced soldier, in defense of homeland security, will become the very emblem of resistance to fluidity. Where soldiers of the past were weakened by "a lack of integration" at the level of the hardware-equipped body (53), the nanosoldier will embody integrated wholeness, defending against "lack" or "disintegration" everywhere. The nanosoldier will resist disintegration by closing off porosity: "The soldier's [nanotech-enhanced] uniform will immediately start to defend against attack by becoming nonporous and filtering the air supply" (48). The nanosoldier's material surface will block everything gaseous or fluidic: "Such a uniform is not only possible, but likely. Nanotechnology-enhanced fabrics that are totally resistant to the penetration of liquids (and are thus stain-resistant) are already made by companies such as Nano-Tex, whose fabric is used in khaki pants sold by Eddie Bauer and others" (44). Nano-enhancement renders both soldiers and civilians "totally resistant to the penetration of liquids" and other destabilizing forces of fragmentation, making them resistant to "stains" at the level of corporeal surface. Indeed, the nanosoldier (and civilians decked out in similar nanomaterials) would appear resistant as such thanks to the absolute erection of the armored body, which hardens against chaos on the outside while containing all vital flows on the inside—a body that becomes the signifier of insemination:

> Researchers at MIT [Institute for Soldier Nanotechnologies] are . . . explor-ing the potential of making a uniform that not only reacts to chemical or biological toxins, but can stiffen and act as armor against ballistic threats such as bullets and fragmentation. . . . The rigid fibers could offer signifi-cant protection against ballistic threats and the armor could be activated or deactivated for convenience, depending on whether the soldier is in battle or just patrolling. . . . The fabric itself could contain a smart ma-terial liner that would adhere to the wound [should something penetrate the armor] and act as a temporary bandage. (52)

The nanosoldier's bodily surface stiffens against threats, containing and bandaging any hemorrhaging of its interior. Detumescing for the sake of "patrolling" or strolling around town, this nano-armor leaves the soldier ready to spring into rigid action according to desire: a sort of nano-Viagra for the tremulous body.

If the body of the nanosoldier and even the nanocivilian could be outfitted to grow erect against fragmentation, fluids, gasses, and all manifestations of molecular abjection, then why not base our entire social structure, our en-tire urban architecture, on the rigidifying function of nanomaterials? Ballis-tic assaults, internal explosions, gasses, toxic fluids, and even tremblings of the soil destabilize our physical and social bedrock; all these manifestations of "matters spiraling out of control" demand, then, the resistance of nano-technology. "How can office and apartment buildings be hardened to reduce the impact of such attacks?" the authors ask. The answer should by now be obvious: "Nanotechnology offers several solutions that could be integrated into the design of future, attack-resistant buildings" (67).

The nano-integration of our living spaces and our cities would suture all gaps, cracks, and pores, all weak spots in our existing strategies for home-land defense. With nano, we could fabricate buildings "designed to be strong but also flexible. Some designs utilize a flexible central mast to support the rest of the building. Constructing these masts of nanocomposite materials or using nanocomposites to replace today's ubiquitous structural steel and reinforced concrete could significantly improve buildings' performance and protection" (69–70). The nanocomposite "mast," the "strong but flexible" central rod supporting the whole structure, becomes the source of "improving . . . performance and protection." Nanocomposite material would appear as aphrodisiac and prophylactic all in one.

Nanocomposite buildings can have massive erections—they can in fact *become* massive erections—on command, or in response to threats of dissemination, even preventing autofragmentation from explosive force (e.g., *jouissance*): "[Nanocomposite] material could vastly outperform conventional construction methods, and a building based on it could withstand a much stronger explosion than anything that exists today" (70). Nanocomposite erectile tissue could therefore be used everywhere—"These same kinds of hardening processes could also be implemented for key infrastructure like water, electricity, subways, mail centers, and communications" (72)—and could integrate everything and everyone, knit us all together and firmly ground us to the soil with the orthotic mast of nanotechnology.

The universal erection of the homeland against internal and external horrors of flowing matters and masses—evidence of molecular abjection at work—would encompass our cities, our nanowarriors, and every civilian who begins *becoming-solider*, it seems, in the very act of putting on "khaki nanopants." In the security crisis of the present age, nanotechnology inflates and grows turgid to overcome any cracks in the armor of national and personal defense. It should be noted that the same rhetoric also historically informed the psychology of the protofascist soldier male in interwar Europe; as Theweleit writes: "Now, when we ask how that man [the soldier male] keeps the threat of the Red flood of revolution away from his body, we find the same movement of stiffening, of closing himself off to form a 'discrete entity.' He defends himself with a kind of sustained erection of his whole body, of whole cities, of whole troop units."[103] So we might say that our so-called new weapons for new wars are really just redeployments of very old weapons for very old wars.

DOMESTIC SECURITY

In 2003 the General Electric Company (GE) launched a television advertising campaign, "Beauty and Brains" (dir. Joe Pytka, 2003), featuring a "mismatched romance" between a scientist and a fashion model. This tale of "the perfect marriage of beauty and brains" allegorizes the combination of aesthetic design and sophisticated technological performance in GE washers and appliances. Though the commercial lasts only sixty seconds and the word "nanotechnology" appears only once and almost invisibly, within the semiotic fabric of the text, nano grows to enormous symbolic proportions. Through its reliance on timeworn traditions of gendering the mind as male and the body as female, this ad manages to fabricate nanotechnology as the nuclear center of the family of the future, the very essence of domestic security.[104]

Considering that GE under the direction of CEO Jeffrey Immelt has become a major industrial player in nanoresearch, this advertisement seems to perform a forward-looking, interventionist function as much as it might help to sell any products in the present.[105] Because GE is, after all, investing in the nanofuture. Margaret Blohm, as director of nanotech R&D at the GE Global Research Center in Niskayuna, New York, said in 2003: "The focus of GE's research is shifting to longer-term projects to create the technologies needed for future markets. . . . Affected GE businesses will be plastics, specialty materials and various product introductions. Nanotechnology will create the second industrial revolution."[106] GE is betting on this nano revolution, and Blohm confesses: "I lose sleep at night because expectations are so high."[107] So amid all this anxiety, the GE marketing department hires the BBDO advertising agency and Pytka Productions to craft a consumerist fable of the happy meeting and "mismatched romance" of nanotechnology with laundry, science and everyday life, machines and the American dream.

The advertisement recounts a story of the love between "Beauty" (Yamila Diaz-Rahi, playing a fictive version of herself)—"Yamila Diaz-Rahi, Supermodel"—and "Brains"—"Dr. Nathan Parker, Professor, Nanotechnology" (figures 25 and 26). The characters of Beauty and Brains unfold all at once through these two images that layer various informational elements into single frames of video. The self-referential content of the images immediately displays the Lacanian logic of their organization, establishing precise relations of body and signifier to gendered subject positions. These images, as the narrator of the ad (voiced-over by Alec Baldwin) explains, are "mismatched"; they do not equal each other, there is something that sticks out as a remainder when they are subtracted, something that one has which the other does not, but which acts as the copula, the equal sign allowing their mismatch to be a match. This something extra—making the two unequal when separate, yet permitting them to come together and connect—would link brains to beauty, and enable love.

The "Beauty" is strategically reinforced as a pure signifier: she "is" beauty itself, and the televisual image of her body is overlapped by other signifiers that reinscribe her status as nothing but a body image: her proper name refers to what we can see—Diaz-Rahi plays herself—and the signifier of her

Yamila Diaz-Rahi
Supermodel

Dr. Nathan Parker
Professor, Nanotechnology

25. The Beauty. ("Beauty and Brains," BBDO/Pytka, 2003)
26. The Brains. ("Beauty and Brains," BBDO/Pytka, 2003)

occupation, the signifier that she occupies, is not just "model" but indeed "supermodel," moreover a "famous supermodel," a *model* model famous in the media landscape for being precisely what she appears to be, a simulacrum of herself. Beauty here is fashioned as pure surface—she displays only what she is—an empty signifier, the aesthetic without substance.

He, on the other hand, is "Brains," and the essence of what he "is" cannot be seen in the ad but projects off-screen, beyond the surfacing effects of the image. His body, his name, and his occupation do not refer to endless simulacra but instead reinforce an inaccessible interior of pure intellect. As "brains," instead of ephemeral beauty, he is abstracted by those signifiers comprising his video frame. But where almost all the signifiers in these analogous video stills are resolutely self-referential—the names, occupations, and body images point indexically to the characters—one word obtrudes as having no referent here, no content within the context of the ad, and certainly no structural parallel in the video frame of the woman whose occupation is "Supermodel." This extra signifier is "Nanotechnology."

Inscribed above the sign of "Professor, Nanotechnology," Dr. Nathan Parker is identified with an extra signifier that sticks out: the thing that he "professes," that he proffers, that he holds out for inspection, is the significant remainder left over when those elements that index his character—his body image, name, and profession—are subtracted from those structurally analogous for the supermodel. She professes only herself, for her profession is what she is: beauty, the model of a model, a supermodel who exists only as the profession of surface, a signifier for another signifier. While he professes himself as a professor, a brain that can project discourse out from itself, the origin and the end of signification, the cultural authority ("Dr.") entitled to profess something beyond himself, namely, "nanotechnology." There is no content to this extra signifier; it is meaningless within the context of this ad, and as a pure signifier, this thing that Dr. Nathan Parker professes becomes an open simulacrum of the same order as the figure of the "Beauty." Both the "supermodel" and "nanotechnology" in this ad are emptied signifiers, available for (fantasy) projection. And here we must think of Lacan's formulation of the phallus itself as the "pure signifier," as the symbol of lack as such, "a unique unit of being which, by its very nature, is the symbol of but an absence," and the constitution of gendered subject positions within the symbolic field as the difference between "being" and "having," that is, between one who "is the phallus" and one who "has the phallus."[108] This advertisement, recessing the male behind his signifiers as the essence informing them, while embodying the female as nothing but signifiers, reinscribes the semiotics of binary gender. In the process, "nanotechnology" is rendered a remainder, jutting out between them, a signifier without signified, professed by the male toward the female as signification of the desire of the other. In this "perfect marriage of beauty and brains," what comes between them is "nanotechnology": the scientist's phallus. ("Nothing 'nano' about it.")

"What can you say about a seemingly mismatched romance between a hardwired computer geek and a famous supermodel? That it was destiny? That love is blind? How about the fact the marriage of beauty and brains can sometimes produce remarkable results? Like the high-tech Profile™ series of appliances from GE." These opening lines of the ad work to reinforce the male as presence and the female as absence, making her a hole in the text and him a hard essence somewhere behind it. First of all, the advertisement herein establishes its blatant intertextuality with the film *Love Story* (dir. Arthur Hiller, 1970), not only appropriating the famously haunting musical score from the film as its own soundtrack but also crisply echoing the film's equally famous opening lines: "What can you say about a twenty-five-year-old girl who died? That she was beautiful and brilliant? That she loved Mozart? Bach? The Beatles? And me." At its outset, then, the advertisement overdetermines the woman as the already deceased love object whose tragic absence reaffirms male presence. In evoking the lines from *Love Story* only to repress their most significant aspect—that the death of the woman has been announced even before the "love story" has begun—the advertisement not only marks the supermodel as destined for erasure, but even erases notice of her death. This hyper-absencing of the woman occurs in juxtaposition to the hyper-presencing of the male scientist. His "hardwiredness" suggests that he is resolute, he does not change, his "being" is fixed in the hardware, there is no fluidity or changeability in him. Whatever happens on the surface does not affect the hard essence inside.

As the professor of 'nanotechnology," he is in charge of the chaotic world of empty signifiers; he would thus also be in charge of her. Significantly, the courtship gift he gives her is the *Dungeons & Dragons Player's Handbook*, a role-playing game manual that, while suggestive of his geekiness, also works to place her in the position, yet again, of simulacrum. For in this gift of rules for role-playing, he would see her behave as a simulated figure, he would see her as representing a fictional role—he would, perhaps, see her as what she is, a supermodel who always plays herself—and this perception of her role as fictive vessel contrasts to the position he presumably reserves for himself, namely, that of the "Dungeon Master." He is the brains *behind* the beauty; he is the presence informing and ontologizing the ephemeral malleability of a model who is nothing but a body, nothing but *matter*.

In fashioning the role of the nanoscientist as the hardwired, present, and living master of simulacra and simulations—master of computers, of role-playing, of the supermodel, and of nanotechnology—this advertisement counteracts other representations of feminized matter, and feminized nanotechnology, as something that could escape and disseminate. The scientific figure here postures as if nanotech were *firmly attached* to the brains controlling it. As such, nanotechnology is made an instrument for insemination, the medium that enables a "geek" to get a "supermodel"—and the medium for control of matter itself.

At the center of this fable is an off-screen act of insemination that produces a child—a child that functions as an allegory of GE appliances themselves, for the child represents the technological product of the "marriage between brains and beauty." The child, now held *between* father and mother (figure 27), is a copulatory signifier linking them—structurally analogous to "nanotechnology," and also analogous to the GE washing machines shown situated *between* man and woman: the home appliances that, like the child, are placed in the very center of the nuclear family (figure 28).

The image shows successful insemination: the delivery of the father's "brains" into the mother's "beauty." After all, the ad makes clear that insemination is a one-way flow, a vector from the father who professes into the mother who receives and nurtures that which gets professed. The child himself suggests that the supermodel is simply a vessel rather than an active agent contributing equally to the product of this mismatched equation. For the boy is clearly his father's son: he wears glasses—professing the sign of geekiness and the brains behind it—and the same style of clothes, but dis-

27. Nano nuclear family. ("Beauty and Brains," BBDO/Pytka, 2003.)
28. Mommy, Daddy, me . . . and the machines. ("Beauty and Brains," BBDO/Pytka, 2003)

closes no analogous traces of the supermodel. She has been reduced to the role of *chora*, delivering the father's son as a photocopy of himself.

But the glasses worn by the child are absurd—what infant would need glasses, or know if he did? Nonetheless, they are profoundly suggestive. They imply that the child is the *legitimate* son of the father and that he has inherited the father's brains. The profession, the output of the father, has routed his abstracted cerebral powers through the female body back into the world. But the glasses also imply that the child has inherited from his father an equally essential deficiency, a primal lack here figured as the loss of visual perception. Quite evidently, the child has inherited the father's *blindness*—a blindness that goes hand in hand with his brains, a blindness that is the very condition that makes mastery possible.

Indeed, this advertisement focuses on a *constitutive* blindness at the center of the domestic romance: as the narrator's voice says, this story is all about the proverbial notion "that love is blind." This cultural fantasy allows a geeky scientist and a supermodel to come together in marriage, for love is here about the inability to see, the desire for the signifier that is never itself but only what it is perceived to be. At the same time, this story is also about the blindness that is actively covered up by appropriation of the technoscientific phallus.

The glasses are the prosthetic instrument whose acquisition would work toward the end of inadequacy, in that having the glasses means the male can still see even when covering up a more originary blindness. An originary blindness that the father passes on to the son, a blindness which, the advertisement implies, can only be overcome by mastering technology, by *taking technology in hand*. In righting his glasses as they slip from his face, putting them in their proper place, getting his technology in hand, the nanotech-

29. With technology in its proper place, love is no longer blind. ("Beauty and Brains," BBDO/Pytka, 2003)

nologist can see the woman as model, as pure image, and he displays his prosthetic gaze for her. In response to his mediated look, she looks back and sees her own desire (figure 29). Thanks to *technology put in its proper place*, this love is no longer blind; rather, it penetrates.[109] In this, the ad follows a certain convention of classical narrative cinema, which, as Laura Mulvey has famously argued, subordinates the view of the camera and the view of the audience to the diegetic perspective of the central male figure on-screen.[110] The view of the camera is thereby functionally rendered a male gaze, with the image attendant to its pleasure. The montage of "Beauty and Brains" similarly makes us aware of the technological mediation involved in visual pleasure, for the look through the lenses of the glasses and the look through the lens of the camera both focus according to the desire of the male subject, the bearer of the look, the bearer of technology itself.

Technology in general, and nanotechnology specifically, becomes the signifier of potency and virility, allowing the ad to end with a tableau of the nuclear family of the future, safe in domestic tranquillity and the contentment available to consumers in the world of late capitalism: a nanoscientist father holds his congenitally blinded son in his arms while an empty mother stands by doing laundry, keeping the family running through household appliances.

But if the mother's death is already implicated by the ad's self-connection to *Love Story*, her relative uselessness and *disposability* for the nuclear family of the future are similarly suggested by the fact that her main task in this household is doing laundry. Yet even the baby can run the washing machine! Indeed, it would seem that this "perfect" machine could even run itself. Mom is displaced entirely by the fantasy of male autogenesis, where men and their reliable technologies have now made the technology of "woman" obsolete. In this as well, the advertisement relies on a certain familiarity with the plot of *Love Story*, for in that film it is woman's fault that the lovers cannot have a baby: woman is a technology that "malfunctions." The young husband, Oliver, asks his family doctor: "Do you know whose fault it is? . . . Two twenty-four-year-olds can't seem to make a baby, obviously one of us is malfunctioning. Who?" Assuredly, it is wife Jenny's fault—not only is she unable to conceive a baby, but her entire body is failing in a massive system shutdown: "Jenny is very sick. . . . She's dying." The catastrophic failure of the nuclear family in *Love Story* is due to a "malfunctioning" woman. So if the GE advertisement's reworking of *Love Story* for the nanotech age reminds us that woman is a technology that mysteriously malfunctions, we can be reassured by the ad that men possess technology that surely works. Technology, of course, like household appliances: "Quite possibly the smartest and most beautiful we've ever designed. Profile™ appliances from GE: the perfect marriage of beauty and brains."

The nuclear family is fulfilled and secured by the signifiers of technology all around them. Even the washing machine—a complicated device, but one that even a (male) baby can master—anchors our domestic architecture, suturing any gaps in the image of the family unit, drawing "Mommy, Daddy, and me" together in a secured dwelling of servile technology. The washing machine is no more a threat to the nuclear family than is nanotechnology; on the contrary, they are now the very instruments that guarantee family values. Indeed, while Mommy loads the washing machine, it is the baby who actually controls the device and determines its operations. Having inherited technical mastery from his father, the infant programs the machine and sets it to work with all the confidence of a subject who, we see with startling clarity, has the technoscientific phallus well in hand (figure 30).

RESISTANCES OF NANOTECHNOLOGY

But must we resist? And, if so, nanotechnology?

For if resistances to nanotechnology are themselves symptoms of technocracy faced with its own uncertainties, humanizing responses to what would otherwise be radically inhuman, psychological operations that would seek to contain whatever revolting threat nanotechnology might represent to itself and to us, then what would happen if we failed to uphold these human (all-too-human) surface tensions?

This question is posed by *Bloom*'s female nanoscientist, Renata Baucum, who recognizes the domesticating function of news reportage on the salvific

30. The technoscientific phallus is well in hand.
("Beauty and Brains," BBDO/Pytka, 2003)

expedition to "seed" and subdue the Mycosystem. She says to Strasheim, the expedition reporter: "That's my problem with you. Thousands of people are going to think whatever you tell them to think. That everything is safe, that changes to the status quo are unwelcome and dangerous?" (47). She deplores the way people blind themselves to what would be radically exterior to them, radically inhuman, accommodating the alienating impact of nanotechnology to psychological defenses: "Such blindness may well be the death of us all" (49). According to Baucum, we blind ourselves to the evidence of alterity; we ignore the danger to our imaginary borders that nanotechnological exteriority would otherwise represent:

> We look at the Mycosystem and we see "scary goo," and never mind about the information content of the spores we see sifting through it every second of every day. We call it "mindless" without ever once attempting to interpret its signals, and when we see *unambiguous signs* of purposeful activity, we simply fail to integrate it. It doesn't jibe with what we know, so we do our best to ignore it. (224)

In our observations of nanotechnology, then, our humanizing tendencies permit us to see nothing except "abject horror" (227).

But if molecular abjection results from presuming an impossibility or a prohibition, a blindness and self-enclosure against the outside and the other, a division of science from science fiction—a resistance, that is, to nanotechnology found within itself—what would happen if we opened our eyes, or touched with our fingers, or tunneled to discover the "impossible," and therefore failed to affirm this prohibition? What if we failed to follow the injunction to preserve these boundaries? If we instead allowed this humanist injunction to dissolve and, in dissolving, also dissolve abjection? We might find ourselves in touch with the outside, with the environment and the exterior, which no longer need be found human or even biological. Renata Baucum again: "There used to be human beings who found importance, even sacredness, in literally everything around them. Not just the living things, either" (223).

This dissolve, or bleed, or stripping of skin would entail an apocalypse, of course, an apocalypse of the human and an end to what we know as "our world." But it would be the apocalypse that Kristeva has called "laughter," an apocalyptic laughter of "horror close to ecstasy."[111] An *impurification* of abjection that eradicates the sublime, those wonders and horrors produced only across a limit, but an impurification that nevertheless preserves the outside in finding "importance, even sacredness, in literally everything around"— a finding of the outside in affirmative ecstasy that, as such, would be the opening to a molecular eroticism and a transfinite laughter: "A laughing apocalypse is an apocalypse without god."[112] A different kind of eschatology, foreseen through nanovision, that extends across humanism and invites the beyond. A radically posthuman apocalypse no longer devoted to a final ordering, a cleansing, righting, or constrainment of matter, the insemination of flows of goo, the oppression of monstrosity . . . no longer devoted to delimiting the boundaries of the possible. Indeed, it would give rise to an age characterized less by symbolic structures and ontic barriers than by operations of the formless and volatile bodies.[113] A new future of process and becoming. A world of difference, of polymorphing and mutability.[114] An era of bodies without organs, matter without form, and nanotechnology without domesticity.

An era of splatter . . .

4

NANO/SPLATTER:

Disintegrating the Postbiological Body

> This much we know; that it isn't simply our lives at
> stake, but the very biology that supports them.
> —Wil McCarthy, *Bloom*

> That's it, man . . . Game over, man! Game over!
> —*Aliens*

At the intersection of nanotechnology and biotechnology, in the hybrid frontiers of "nanobiology," we once again find the future. The future of technoscience and socioeconomic development, certainly—the administration of
the U.S. National Nanotechnology Initiative, for one, sees nanobiology as
an essential route for advancing nanoscience and its attendant industrial
potentials—but even the future of "life" itself and what counts as living
in the nanotech era.[1] For through the perspective of nanovision, not only
does the material horizon of the organism shift from the microscale to the
nanoscale (recent claims about "nanobacteria" and "minimalist organisms,"
for example, instantiate this shift), but life ceases to remain fixed in the domain defined by prevailing conceptions of "biology."[2] As the nanotheorist
Charles Ostman suggests: "The very definition of life itself may be perched
at the edge of the next great revolution in medicine—nanobiology. What is
emerging now are technologies and applications in the arenas of biomolecular 'components' integrated into microscale systems, . . . synthetically engineered quasi-viral components, modified DNA and related pseudoproteins,
biomolecular prosthetics, and biomolecular organelle component 'entities'

. . . [which] will redefine the very essence of what is commonly referred to as 'life.'"[3]

With repeated emphasis on the consequences for "life itself," Ostman envisions an inevitable global transformation in the wake of nanobiology, writing that the "'history of the future' is already unfolding and the primary elements of this evolutionary eventstream that are poised to reshape the economies of the world, and perhaps even the very definition of life itself, are currently at hand."[4] Like other speculative launches of nanovision, this collapse of science-fictional speculation into the technological present involves much more than hype. For Ostman's discovery of the "already unfolding" future and traces of technological Singularity at the site of nanobiology works performatively to relocate life elsewhere, beyond biology. The "biomolecular components" of engineered microsystems already become present examples of the "redefinition of life" yet to come: they are postbiological creatures in sufferance, provisional "entities" displaced prematurely from the "evolutionary eventstream." Indeed, nanovision regularly fabricates autonomous life-bearing agency outside traditional topographies of organismics, cytology, and genetics. Under the scrutiny of nanovision, biology as such transforms into the "newly emerging arena of nanobiology, in which the molecular components of living organisms can be 'disassembled' and reconstructed to create viral-like entities" and "biomedical 'systems' which . . . mimic the physiology of living organisms in their operational and behavioral characteristics."[5]

In this "arena," vital signs migrate beyond biological space via a textual procedure that involves a promise, a vision of organelle components liberated from the biological organism, of cytoplasmic systems spilled from phospholipid membranes and thereupon enabled to self-actualize. It is a narratival act of opening bodies through nanotechnological processes and reabsorbing subcellular molecularities into the epistemic domain of nanobiology, the domain of their becoming-entity.

The molecular "entity" materializes as an object of technoscientific knowledge within a narrative discourse of corporeal violence immanent to nanobiology. This narrativity—which takes place within what I will call the "scene of disintegration"—deterritorializes the components of the body and simultaneously destines the molecular machines of the living cell toward a future where "life itself" has been "reshaped." The scene of disintegration therefore operates through the specular and speculative movements of nanovision: it makes the nanoscale symbolically accessible by "disassembling" organisms and translating molecules into technoscientific representations, but in doing so it also visualizes the "history of the future" and reverse-transcribes the destiny of molecular entities into the present.[6] As Ostman suggests, the radical reshaping of life in the nanofuture is "an emergent, transformative phenomenon which has already become manifest."[7]

Disintegration represents a narrative form of experimentation, both a way of imagining nanobiological operations in advance of their performance and a rhetorical simulation that plays out abstractly, even when they are performed, as their script or tropic protocol. Like "exploratory engineering" in nanotech broadly, disintegration exemplifies what Nicolas Pethes calls the "exploring" program that establishes common praxis for "both scientific and literary modes of experimenting: exploring possible forms of knowledge."[8] Pethes suggests that the "fictive experiment[s]" of popular literature, such as those staged in the medical thriller, are able to "test nonexistent ways of knowing and their effects in the extension of a narrative, insofar as the narrative provides the temporalization and causal suppositions an experiment requires." I would extend this argument to the science (fiction) experiments of speculative nanobiology, suggesting that the narratives of disintegration which inform both nanoscience and nanofiction work not only to explore "nonexistent ways of knowing" but also to produce them, to make them exist. In tracing the narrative operations of nanovisionary disintegration, I will argue that they work even now to create the conditions of possibility for the molecular machines of the body to be released into the world as self-reproducing autonomous agents. In other words, the science (fiction) experiments of nanovision engineer the epistemic space needed to accommodate molecular forms of "postbiological life."

Defined by the complexity theorist Stuart Kauffman, the "molecular autonomous agent is a self-reproducing molecular system able to carry out one or more thermodynamic work cycles," and as such would constitute a new "proper definition of life itself."[9] Accordingly, John Shirley's novel *Crawlers* (2003) describes the kind of molecular agent fabricated by nanobiology as "nanotech life."[10] Kathleen Ann Goonan's *Queen City Jazz* (1994) calls it "Enlivenment," for it enhances otherwise inert matter with an uncanny vitality.[11] Wil McCarthy's *Bloom* (1998) calls it "technogenic life," and while certainly not biological, as far as "life" goes, "TGL [technogenic life] is the real thing; it eats, sorts, metabolizes, reproduces" (100). This "real thing" imagined by science and science fiction is already anticipated within the representational space of the experimental laboratory: for some nanoscientists, the belated arrival of technogenic entities in the future would merely complete a process already activated by the theoretical machinations of nanobiology, the redefinition of "life itself" in light of the subcellular agencies released from biology to become "molecular self-assembling subcomponents organizing as self-modifying organelle entities, which . . . [will] become ubiquitous, and flourish on the physical terraform."[12]

Traversing related scientific fields like artificial life, evolutionary robotics, and prebiotic molecular evolution—those fields working theoretically or experimentally to evolve something like life "from the bottom up"[13]—nanobiology presupposes an evolutionary trajectory of technogenic life stretched from the prebiotic into the postbiotic, into a future where biomorphs must learn to coexist with technomorphs. The researcher Steen Rasmussen and his colleagues have written that nanobiology will "eventually produce dramatic new technologies, such as self-repairing and self-replicating nanomachines. With metabolisms and genetics unlike those of existing organisms, such machines would literally form the basis of a living technology possessing powerful capabilities and raising important social and ethical implications."[14] Into this nanobiological future, as yet unarmed with an adequate posthuman ethics to accommodate these new "living technologies," these new forms of "machinic life," we are encouraged to proceed with "cautious courage."[15]

But this is a future that does not wait for the future—it happens now, and already. Recent experiments by nanobiologists like Nadrian Seeman and Bernard Yurke, using DNA as a self-assembling engineering material for three-dimensional structures and molecular motors, suggest the nascence of "elaborate robotic functions" at the nanoscale and point to a "nanotechnological dream machine . . . that can replicate": "It is still a far cry from the replication achieved by every living organism, but by the time the Watson-Crick centenary comes around, we should have DNA-based machines that do as well."[16] Such automorphic DNA machines, extracted from cytoplasmic contexts and the realm of biology entirely, self-organizing and self-replicating "as well as every living organism" but embodying a machinic alternative to biotic systems, contain the future in themselves: "Human-made nanomachines that are powered by materials taken from living cells are a reality today. It won't be long before more and more of the cells' working parts are drafted into the service of human-made nanomachines. As the merging of living-nano and non-living nano becomes more common, the idea of *self-replicating* nanomachines seems less and less like a 'futurist's daydream.'"[17] For the nanoscientists Robert Freitas and Ralph Merkle, these dreams and dream machines have almost become indistinguishable from waking life: "Self-replication is a hallmark, though no longer the exclusive province, of living systems. . . . The technology presently exists to create artificial self-replicating hardware entities."[18] And with self-replication comes the potential for evolution . . . the evolution of machinic life beyond the biological world.[19]

Through these providential signs of things to come, the nanomachines of the body are already released into the promise of the future—not as a technological event, but as a semiotic one. The natural biotechnology of the cell has been *prefabricated* as an embryonic stage of technogenic life by the speculative gaze of nanovision, awaiting only deterritorialization (or disintegration) to be sent on its way to becoming-entity. As Gilles Deleuze and Félix Guattari have written, "Only something deterritorialized is capable of reproducing itself"[20]—that is, the degree to which it is deterritorialized measures its ability to self-replicate independently from the production or maintenance of higher-level formal structures or aggregates—and at the moment the autonomous molecularity escapes the enveloping cellular organism to become self-assembling, self-replicating, and self-organizing, biology's historical monopoly on life has come to a close. The newly born nanomachine no longer needs the support of biology: it does not need a "master molecule" or a "genetic program" to coordinate and control it;[21] it does not need a homeostatic or autopoietic system to preserve it;[22] it does not need any crutch of biologism at all.

We see in nanobiology, then, a resonant coordinate of what Richard Doyle has called the "postvital" era, where life disappears from the closed interior of organisms and dissolves into molecular informational processes, codes, feedback systems, and programs, but returns, undead, as the absent origin of those very scientific practices—from molecular biology to genomics to artificial life—effecting its dislocation.[23] It is not so much that biology has been destroyed by these postvital sciences but that the biological now manifests "beyond living," a sublimation of life in excess of its reduction to technologies of information. For Eugene Thacker, this postvital return of the biological beyond its infomechanical reduction is an effect of "biomedia," the "technical recontextualization of biological components and processes"—exemplified in fields like bioinformatics, systems biology, biocomputing, and nanomedicine—that ultimately "returns to the biological in a spiral, in which the biological is not effaced . . . but in which the biological is optimized, impelled to realize, to rematerialize, a biology beyond itself."[24] As we will see, disintegration's multiple remediations of the "technological" within the "biological," its serial enframing of the machine both within and without the body, effectively projects machinic life as the biology beyond biology, the afterlife of life. Or to put it another way, disintegration unleashes the machinic phylum.

The machinic phylum, as Deleuze and Guattari have written, is "matter in movement, in flux, in variation, matter as a conveyor of singularities and traits of expression."[25] The machinic phylum describes all the self-organizing processes in nature. It materializes in the spontaneous assemblage of cooperating populations of elements (atoms, molecules, cells, organisms) and operates as intensifications, accelerations, sedimentations, and destratifications of the flows of matter-energy-information. The nanoscientist Jean-Marie Lehn has written of the machinic phylum: "From divided to condensed and on to organized, living, and thinking matter, the path is toward an increase in complexity through self-organization. . . . Self-organization is the driving force that led up to the evolution of the biological world from inanimate matter."[26] Traversing the organic, mineral, and social worlds, the machinic phylum therefore gives rise to a new conceptuality of the meaning of life, the recognition of "nonorganic life"—"a life proper to matter, a vital state of matter as such, a material vitalism"—in the creative expressions of the physical world.[27] Tracking evolutions of the machinic phylum across biological, chemical, cultural, linguistic, and geological history, Manuel De Landa writes: "In a very real sense, reality is a *single matter-energy* undergoing phase transitions of various kinds. . . . Rocks and wind, germs and words, are all different manifestations of this dynamic material reality . . . different ways in which this single matter-energy *expresses* itself."[28]

This expressivity of the machinic phylum—especially the self-organizing performance of the molecular machines already inside the body—in itself suggests to nanoscientists the way to bring the nanofuture into being, for "self-organization processes . . . provide an original approach to nanoscience and nanotechnology,"[29] and "molecular self-assembly . . . [which] is centrally important in life . . . seems to offer one of the most general strategies now available for generating nanostructures."[30] But if the machinic phylum thus "provides" for nanoscience and "offers" its services, then the extent to which nanotechnology can be considered the "total, absolute control of matter" must once again seriously be questioned.[31] B. C. Crandall ventures that nanotechnology is "simply a descriptive term for a particular state of our species' control of materiality," a step in our species-destining toward "absolute control."[32] Yet having provided both anticipation and means for matter to become self-capacitating, nanotechnology may actually be in the process of demolishing the anthropic concept of control entirely.

According to Ted Sargent, "Nanotechnology is coordinated movement, a choreographed dance among atoms and molecules to achieve a desired effect. It harmonizes within Nature's own set of rules to coax matter to assemble into new forms."[33] Already reliant on the self-organizing performativity of the machinic phylum for many of its assembly techniques, nanotechnology tentatively opens to what Andrew Pickering names the "dance of agency" between human and nonhuman entities in the "posthumanist space" of technoscience.[34] It begins to acknowledge, and speak for, what Bruno Latour has called the "Parliament of Things" in its own laboratory environment.[35] For some nanoscientists, certainly, the self-organizing rhythms of the machinic phylum seem to have evolved human life and human science precisely to become the parliamentary representative—or choreographer—of matter itself. As Lehn suggests, "How matter is becoming and has become complex is the most fundamental question raised to science, for it indeed asks how (and why?) the evolution of the universe has given rise to an organism capable of asking this very question and of generating the means to answer it by creating science."[36] Nanoscience thus tangos with the machinic phylum; they respond to each other, coaxing each other into new forms, moving harmoniously together across endless frontiers. And if this dance is already becoming evident in contemporary laboratory life, it is a mere trace of the imaginable future in which a "living technology" has been liberated from within our own bodies.

While the scientific accounts of nanobiology and the nanotech novels that I analyze in this chapter differ in many respects, they coincide in the common practice of staging and restaging disintegration, of fabricating the performative space for sending, posting, or destining the molecular components of the biological body into the postbiological future. This nanovisionary movement, I argue, begins laying the foundations for an ethics of human-technology relations that eschews power disparities between biological molarities and postvital multiplicities, wholes and parts, bodies and nanoscopic machines. For within the future of self-capacitating matter made thinkable by nanobiology, the human ("our species") as a metaphysical construct of territorialization—stripped of its species-destining toward mastery—dissolves into a posthuman network of distributed agencies where, as N. Katherine Hayles has said, "a dynamic partnership between humans and intelligent machines replaces the liberal humanist subject's manifest destiny to dominate and control nature."[37] In this network, we find technogenic life traversing software, hardware, and wetware, generating new machinic-semiotic flows that consume the privileged spot of the biologistic human subject. With the promise of postvital molecularities embedded in the narratival experiments of nanovision, the era marked by the reflexive construction of the "human" and biological "life" vaporizes into history as life in the coming nanotech era explodes, bursts, spreads, and disseminates across striations of the machinic phylum.[38] In a word, life "splatters."

THE SCENE OF DISINTEGRATION

"Splatter," in the vocabulary of literary and cinematic horror, has come to refer to a representational moment in which the human body is violently torn asunder, shredded, sliced, hacked, dismembered, melted, and transformed, splattered as semiotic fluid into ghastly forms of monstrous abjection. As the defining motif of "splatterpunk" fiction—represented by the wettest productions of auteurs such as Clive Barker, Poppy Z. Brite, John Shirley, Edward Lee, George Romero, Lucio Fulci, David Cronenberg, and Peter Jackson—splatter is the figural mechanism through which narratives of "extreme horror" create meaning: in these texts, "mutilation is the message."[39] By disrupting the body's boundaries and the social codes adhering to them, splatter viciously unsettles the economies of corporealization, and Jay McRoy has argued that at the moment of splatter, the "spectacular and graphic deconstruction/transformation of the 'human' form" enacts a radical revision of normative embodiment, suggesting possibilities of somatic experience other than those encountered in the historical accident of human morphology.[40] Or as Judith Halberstam has written, the figures that "emerge triumphant at the gory conclusion of a splatter film are literally posthuman, they punish the limits of the human body and they mark identities as always stitched, sutured, bloody at the seams, and completely beyond the limits and the reaches of an impotent humanism."[41]

Fictive nanonarratives frequently appropriate this horror-image of splatter, this primal scene in which the coherence of the human body is irrevocably fragmented, as a means to vivify the tangible and starkly inhuman impact that nanoscience is beginning already to have on our lives. While not all nanofictions feature splatter imagery—many texts focus on "dry" (nonbiological) nanotechnology, and their aesthetic is correspondingly desiccated— scenes of splatter appear frequently enough to be considered a major topos of the genre. Across multiple media, these stories revel in gruesome liquidation. In novels like Dean Koontz's *Midnight* (1989) or Richard Calder's *Dead Girls* (1992) and its sequels, nanobiomachines spread like viruses and hideously mutate the infected. In films like *Jason X* (dir. James Isaac,

2002)—a late entry in the venerable *Friday the 13th* series—the mass murderer Jason Voorhies morphs into an even more efficient killing machine thanks to "nanotechnological reconstruction" (figure 31): revolutionary science here simply intensifies an ongoing calculus of slaughter, the cinematic algorithm of the "body count." In comic books like *Xombi* (1994–96) and Ben Templesmith's *Singularity 7* (2005), nanites relentlessly "deconstruct" organic bodies in order to reconstruct new bionic ones (figure 4). In role-playing games like GURPS *Ultra-Tech 2* (1999), "splatter nanomachines" circulate through bloodstreams and detonate on command, ripping victims to shreds.[42] In video games like *Nano Breaker* (2005), entire populations are devoured and monstrously reconfigured by nanos whose self-replicating ability relies on extracting the iron from human blood (figure 32). Everywhere in narratives of the nanofuture, the human body appears ripe for destruction—and desperate for an upgrade.

At one level, scenes where nanomachines perpetuate grotesque violence against the flesh simply advertise possible dangers of nanotechnology—especially the abject threat of gray goo and the apocalyptic horrors of global ecophagy. But I want to examine the scene of splatter in nanofiction—the *scene of disintegration*—as the "experimental" narrative procedure whereby molecular agencies become detached from biology and fabricated as postbiological life. I will also present the scene of disintegration as a speculative challenge to body ideals based on hierarchical integration of component parts, offering instead an account of the automorphic organism as a self-organizing collective of "interdependent but independent" molecularities (as *Crawlers* puts it), thereby redistributing the criteria of life beyond biology and across other forms of machinic existence.[43]

In John Robert Marlow's *Nano* (2004), the nanoscientist John Marrek develops an advanced nanotechnology capable of breaking down matter and restructuring it into any form. But corrupt government agents, recognizing the revolutionary effects that this technology would have on every level of human existence—bodily, cultural, economic, political—attempt to kill Marrek to prevent him from revealing his invention. In the extended chase sequences that make up the novel, Marrek releases herds of nanobots against his pursuers, and as the nanobots multiply and swarm, the reporter Jennifer Rayne watches the grisly consequences with rapt fascination:

> What she saw was beyond the experience of any living being. Twenty yards distant, men, guns, and vehicles *crawled* as though covered with some unseen, colorless substance. . . . As she watched, all of these things,

31. *Jason X*: "Evil Gets an Upgrade." Nanotech ups the arsenal of serial killing: flesh collides with hardware, and the bodies hit the floor. (*Jason X*, New Line, 2001)

including the men—*disappeared* from the outside in. The men screamed horribly, dissolving into the nothingness before her eyes—first skin, then muscle and organs, followed by bone. The cars, too, disintegrated before her eyes. So horrified and enthralled was she by the sight of it that she almost failed to notice that the crawling surface covered the street as well—and was spreading rapidly outward in all directions, including hers.[44]

The unprecedented, sublime spectacle of the scene of disintegration—"beyond the experience of any living being"—is both "horrifying" and "enthralling" because it violates human perceptions of the stability of the macroscale world. The viewer is mechanically "riveted by the scene" of people being "taken apart, atom by atom" (108) because it undermines the conceptual integrity of the body, exposing the precarious foundations of human structure: "The screams of the terrified men filled the hall as the calcium-eating nanites entered their bodies through their skin and dissolved their bones from the inside. Bodies collapsed inward and lay on the floor like so many puddles of warm flesh" (319). Nanofiction hyperbolizes the extent to which the body, even in its everydayness, is in a state of constant nanotechnological disintegration, systematically made and unmade by the flows of what becomes, in nanobiology discourse, the "molecular technology" of the organism. As the nanobiologist David S. Goodsell writes, "Even structures that one might feel are permanent, such as bones, are continually disassembled, repaired, and rebuilt" by the "bionanomachinery" always at work inside the body.[45] And if nanomachines liquefy other material assemblages while disintegrating human bodies, as they frequently do in nanofiction, this simply demonstrates that the human is just one possible formation of the fits and eruptions of the machinic phylum. According to *Nano*: "A nanite did not distinguish between materials living and inert," because "any object on earth can be *torn apart* into its constituent atoms, which can then be used to build something else" (199, 139).

Suggesting the debt that the scene of disintegration owes to conventions of horror-splatter, Marlow writes that the nanites are "molecular sledge-hammers, or chain saws" (229), miniature hardware-killers, perhaps akin to the iconic Leatherface from Toby Hooper's *The Texas Chain Saw Massacre* (1974). Like Leatherface—the dispossessed slaughterhouse worker disci-plined to reproduce mechanical violence as the invisible base of late capital-ism, the deranged serial killer who eats his victims after demolishing them with chain saw and sledgehammer—nanites are also mindless recyclers of human parts, consuming "all human flesh in the building" (204) to repro-duce their own ravenous, material flux: "The stricken team leader staggered screaming across the floor, right hand and face disappearing as nanites disas-sembled the outer layers of his brain and enveloped his body," transforming him into nanoware as "each new nanite set about the task of employing each of sixteen arms to rip apart its surroundings and construct new nanites"

32. *Nano Breaker.* Nanomachines disintegrate all the inhabitants of an experimental re-search island, transforming them into Orgamechs, "living mechanical organisms whose bodies are comprised entirely of microscopic machines, from the molecular level on up." (*Nano Breaker*, Konami, 2005)

Every nanomachine on the island malfunctioned;

(199–200). Similarly, in Shirley's *Crawlers*, a population of experimental "nanocells," called "breakouts" or "crawlers" for their nomadic and insidious nature, propagates itself within a small California town through the hostile consumption of bodies: "The nanocells used the body's resources to make more and more of themselves, and took over completely" (325). Nanofiction therefore contributes to what Mark Seltzer has called the "splatter codes" of postindustrial machine culture, the systemic processes of consumption recycling bodies as cybernetic flows of matter and information, the absorptions and relocations of the body into serial repetitions of addictive violence constitutive of postmodernity.[46]

Molecular violence against the body becomes the source of mechanical reproduction, and human meat is rendered the baseline of machine ecology. As envisioned by real science as much as science fiction, nanomachines would seem to have an appetite for our proteins, our nucleotides, our sugars, and our fats. Kurzweil explains: "Living creatures—including humans—would be the primary victims of an exponentially spreading nanobot attack. The principal designs for nanobot construction use carbon as a primary building block. . . . Pathological nanobots would find the Earth's biomass an ideal source of this primary ingredient. Biological entities can also provide stored energy in the form of glucose and ATP."[47] Certainly, engineering diagrams for implantable nanoscale devices even today involve biofuel cells that draw voltage from the glucose-oxygen reactions in human blood. Researchers at the University of Texas, Austin, report that "when fully engineered the simple miniature biofuel cells will be of value in powering small autonomous sensor-transmitter systems in animals and in plants."[48] Similarly, nanoscientists at Panasonic's Nanotechnology Research Laboratory in Japan have developed a device that generates electricity from blood glucose, a "'bio-nano' generator [that] could be used to run devices embedded in the body, or sugar-fed robots."[49] Such intravenous nanomachines, sucking energy from human bodily fluids, have appropriately been nicknamed "vampire bots."[50]

Humans thus become sanguinary batteries for the vampires of the nano-world, raw materials remade in the machines' image. In *Crawlers*, the "breakouts climbed into some people and reorganized them," while "others were just . . . parts" (5), reserved for extraordinary cyborg experiments combining human tissue with mechanical and electronic components, testing "new shapes and formats, looking for different models that worked," "new ways to think life and organizing it" (274). The body in pieces, dispersed among nanobiological modes of production, suggests other possible assemblages, other morphogenies: "[The breakouts] had stripped all the skin off Ahmed's skull, to be used in some other project" (5). Reassembled differently, its fragments inserted into puzzling inhuman projects, the body in extremis becomes bricolage, a pastiche of recombined and recycled detritus mixed by the subcellular sampling artists of the nanofuture.

Through serial disintegrations of bodies, self-reproducing molecular entities find new possibilities for autogenesis, for tinkering and restructuring themselves out of the charnel wreckage, the scraps of an obsolete material organization called "biology." Nanobots rising from the gore: a figuration of machinic evolution from below, from within, from the "bottom up," as nanodiscourse imagines it. Exemplified in Wil McCarthy's *Bloom*, biosplatter forms the condition of narratability of the "technogenic" epoch, for the novel's first paragraph opens with the disintegration of a man from the bottom up—a resonant symbol for the ascension of nanolife: "It started with the feet. These things usually do" (1). A voracious and unstoppable population of self-replicating nanomachines dubbed "mycora" has overspread and consumed Earth as well as the entire inner solar system, forcing human refugees to retreat to the moons of Jupiter. But even in the cold reaches of space, stray mycora find their way into underground human habitats and begin to "bloom," reproducing themselves by dissolving bodies and material structures into "the rainbow mist, the living dust, the bloom of submicroscopic mycora still eating everything in their reach and converting it to more of themselves" (2–3). On the eve of a mission to investigate the Mycosystem—the ecological collective of mycora inhabiting the inner planets—a bloom erupts and devours the mission security captain: "[His] body did not come apart at once into threads and dust, but his skin had gone rainbow-crystalline with mycoric frost before he'd even hit the floor, and of course he never did rise" (56). Indiscriminately, the mycora liquefy organisms and nonorganic structures: "The air vent and the wall it was part of began to boil, their substance turning fluid, turning into rainbow-threaded vapors as the tiny, tiny mycora disassembled them molecule by molecule" (54–55).

During the scene of disintegration, organisms and air vents splatter, and the nanomachine becomes the form of life that emerges precisely through the molecular redistribution of wetware and hardware, the form of life that reproduces itself by desedimenting the machinic phylum, "vaporizing" and "turning fluid" temporary accretions of materiality and bootstrapping itself together out of the resultant goo. *Bloom*'s narrator, the reporter John Strasheim, observes the self-replication of "nanolife" through an electron micrograph:

Not quite crablike, not quite urchinlike, not quite organic in appearance. A tiny machine, like a digger/constructor but smaller than the smallest bacterium, putting copies of itself together with cool precision, building them up out of pieces too small for the micrograph to capture. In short, a pretty typical piece of technogenic life. (6)

Reliquefying the material flows that cut across biological and nonorganic assemblages, disintegration enables endless reconstructive possibilities for "technogenic lebensforms" (95), and the mycora produce complex behaviors in multiple modes of organization: "They're a lot more than just [digesting machines]. Individually, they're complex and mysterious enough. In groups, their behaviors become even more intricate. They communicate back and forth, change state based on information they receive" (139). Thus seemingly random structurations of the Mycosystem—including formations that look suspiciously like human bodies—are really effects of molecular agents working in coordination, as the novel's nanobiologist, Renata Baucum, suggests:

> *Maybe* the Mycosystem generates these structures itself; it's not nearly so homogeneous or random as people seem to think. You know the term "emergent behavior"? Small actions repeated a million times over, with decidedly macroscopic results. Our bodies aren't lumps of undifferentiated flesh, and the Mycosystem is *not* a lump of undifferentiated mycora. (96)

The nanobiologist analogically refigures the human body and the Mycosystem as inhabiting the same plane of machinic organization, both macroscopic emergent structures produced through "small actions repeated a million times over" by the differentiated behaviors of molecular agents. The systemic "body" and its formations, its corporealizations, its vital signs, represent a level of complexity matched entirely by the complexity of its components, and life (the term describing this level of complexity) applies regardless of integration or disintegration:

> "I think [the mycora are] *alive*, yes. I know they are, and in terms of the number of subunits and the linkage between them, the complexity is comparable to ours, to any higher animal's. But for all that, they're just components in a larger system."
>
> "I always figured they were just goop," [Strasheim] said.
>
> "Yeah. So is hemoglobin." (184)

Comparing molecular linkages to mammalian complexity and splattering mycoric "goop" into that stratum of the animal body occupied by hemoglobin (connecting disparate tissues through communal allocations of oxygen), Baucum exposes her disintegrative vision of "larger systems" as collectivities of molecular agents that are themselves "alive." Which is not to say that hemoglobin is thereby "enlivened," but as an analogue to the "interdependent but independent" mycora, hemoglobin is made to stand for the molecularities of the biological realm destined toward the future as technogenic life once released from the hierarchies of the organism.

Hemoglobin, then, is put into "becoming-technogenic life" by the *evolutionary* perspective of *Bloom*, which sees the Mycosystem as an autonomous "entity" (227), constantly evolving alongside its independently evolving components: "It's a devious one, that Mycosystem. . . . Not in a deliberate sense. I mean in an evolutionary one. It's always trying something new. Forms, behaviors, the mycora themselves" (129). Postbiological evolution, the heterogenesis of molecular parts and molar wholes, takes place by virtue of the disintegrations that conceptually open up formerly closed systems, that volatilize bodies as blooms of cooperative molecular entities. Nanofictions, along with Deleuze and Guattari, comprehend these molecular entities as "formative machines . . . engaged in their own assembly, operating by nonlocalizable intercommunications and dispersed localizations, bringing into play processes of temporalization, fragmented formations, and detached parts . . . and where the whole is itself produced alongside the parts, as a part apart."[51] Nanosplatter thus not only marks an end to one conception of the body and its hierarchization, but an end to biologism proper and its maintenance of the organism as the teleomechanical horizon of life. In *Bloom*'s haunting summation: "This much we know; that it isn't simply our lives at stake, but the very biology that supports them" (258).

This anticipatory closure of biologism plays out in Michael Crichton's *Prey*, where a covert scientific institution develops nanobiological agents— autogenic "mechanical organisms"—that interact together as self-organized "swarms" (177). The swarms exhibit intelligent behavior and even predatory tendencies, reproducing themselves through bio-disintegration: "The swarms consume mammalian tissue in order to reproduce" (194). They begin hunting their human creators, forcing the scientists to conclude: "The swarm reproduces, is self-sustaining, learns from experience, has collective intelligence, and can innovate to solve problems. . . . Which means for all practical purposes, it's alive" (176). Not only a new form of life, the swarms also supplement and supplant the biological entirely. Bodies inhabited by the swarms become reconstituted at the cellular level by mimetic nanomachines, as the narrator, the computer scientist Jack Forman, discovers when he triggers an electromagnet to disable the swarms:

> [Julia's] mouth was opened as she screamed, a steady continuous sound, her face rigid with tension. I held her hard. The skin of her face began to shiver, vibrating rapidly. And then her features seemed to grow, to swell as she screamed. I thought her eyes looked frightened. The swelling continued, and began to break up into rivulets, and streams.
>
> And then in a sudden rush, Julia literally disintegrated before my eyes. The skin of her swollen face and body blew away from her in streams of particles, like sand blown off a sand dune. The particles curved away in the arc of the magnetic field toward the sides of the room. (339)

Julia's body has been replaced in its tissues with technogenic creatures. The biological and all its functionalities, even complex properties of behavior and cognition, have been replicated and replaced by nanotech agents working together as a swarming collectivity. The human surface vanishes in a scattering stream of molecular sand, evoking Michel Foucault's resonant prophecy for the historical end of the human era: "One can certainly wager that man would be erased, like a face drawn in sand at the edge of the sea."[52]

The scene of disintegration instantiates a posthuman splatter-narratology whereby the body is broken down into its constituent molecular components, its tissues ruptured, its cells lysed, its organs and organelles spilled into the surrounding environment and thereupon instantly restructured into autonomous nanomachines. Far from being isolated to the imaginary terrains of fiction, nanosplatter informs and performs identically within the discourse of contemporary nanoscience. For example, when the nanobiologist David S. Goodsell tropologically reconstructs the active biomolecules inside the human body as "atomically precise molecule-sized motors, girders, random-access memory, sensors, and a host of other useful mechanisms, all ready to be harnessed by bionanotechnology," he discursively enacts the postvital disintegration of the cytological realm, now "already harnessed" by the transformative power of nanobiology.[53] The biological is torn apart and refabricated as the nanotechnological: "As you read these words, 10,000 different nanomachines are at work inside your body. These are true nanomachines" (9). These nanomachines are released into the world by pulverizing the organism and fracturing the cell: "Remarkably, many of these nanomachines will still perform their atom-sized functions after they are isolated and purified. . . . They do not have to be sequestered safely inside cells. Each one is a self-sufficient molecular machine" (9). The organism represents an obsolete morphogeny, for although "the need to enclose an organism and separate it from the environment is a product of evolutionary selection," developed "to enclose genetic information within a defined cell so that the cell may compete with other cells," once nanobiology dispenses with natural evolution, "the requirement for containment may also become unnecessary" (220). The functionalities of the organism can be replicated and improved by unbounding the nanomachines within: "If we are doing the designing, we don't have to worry about [evolutionary] competition and we can simply build the machine of the best design. One might imagine a 'gray goo' composed . . . of a collection of bionanomachines that come together and perform specific tasks and then disassemble when finished" (220–21). To the nanobiologist, current in vitro research on the self-replication of information-carrying molecules demonstrates progress toward an imminent postbiological future, because even "today, bionanotechnology opens the possibility of creating new life forms" (277). These postbiological entities of the "opened" nanoworld would not be "limited to biological materials, or to physical materials at all" (278), and they could even strive for sentience: "If consciousness also turns out to be reducible to physical principles, creation of consciousness in artificial objects (beings?)" would be completely within the capabilities of nanobiology (298). Artificial life and artificial intelligence and ten thousand "self-sufficient molecular machines" crawl out of the wreckage of the now "unnecessary" organism.

The nanovisual gaze in both science and fiction unleashes the machines of the cell and fosters the ascension of postbiological life by splattering the body. Nanotechnology therefore destines the future through a fundamentally *corporeal* violence, a violence immanent to its own rhetorical practices, a violence toward the embodied present that determines the postvital nano-future as already inevitable. This irreducible violence of nanotechnology's transformative way of scrutinizing the world registers in Marlow's *Nano* when the nanoscientist John Marrek determines that unleashing the post-human nanofuture has become unavoidable—he determines this through a logical process that is itself inherently and clinically violent: "Jen had at first argued against John's plan, but in the end he had dissected and eviscerated each objection with the cold blade of reason" (354). The specular and speculative logic of nanovision, as a way of thinking, a way of seeing, a way of rhetoricizing the real, is itself a *splatter-technology.*

Shared across nanofiction and nanoscience, then, the violent logic of nanosplatter deconstructs the biological body and refabricates it as a contingent and temporary assemblage of self-actualizing nanomachines. For example, in *Prey*, not only do autonomous nanoswarms colonize human bodies, but the disintegration-events in the novel trigger an epistemological revision about the ideational coherence of the biological organism, suggesting that *already* "a human being is actually a giant swarm. Or more precisely, it's a swarm of swarms, because each organ—blood, liver, kidneys—is a separate swarm. What we refer to as a 'body' is really the combination of all these organ swarms. . . . [It's] literally nothing but a swirling mass of cells and atoms, clustered together into smaller swirls of cells and atoms" (260). Similarly in *Bloom*, when we learn that the human beings disintegrated by the mycora have not been destroyed but merely "unpacked," we see that the organism is merely one temporary assemblage that molecular entities can accrete, one form of symbiotic cooperation where multitudes of nanomachines can come together for mutual support. But while the "unpacked" human body can be restructured, it turns out that "technogenic life" prefers other modes of phenomenal materialization besides that of biology: "This limited form, this solid flesh, confines. Very few choose to return to this form, this solid flesh, because it *confines*. We choose to remain Unpacked" (295). Or, as the nanoscientist J. Storrs Hall has written of his desire to be transformed into a nanotech utility fog, "What I want to be when I grow up, is a cloud."[54]

The imagined preference of postbiological life for forms of embodiment beyond the organism even takes a drastic turn *beyond the molecular* in Greg Bear's *Blood Music* (1985), where the end of humanity—the end of biology, the end of life and the universe as we know it—plays out spectacularly through an escalating sequence of disintegrations. The development of nanotechnical "biologic" triggers a serial relocating of life from the organismal to the cellular to the molecular to the quantum mechanical, becoming immanent information at the fundamental possibility structures of spacetime. In Bear's novel, the geneticist Virgil Ulam creates sentient single-celled organisms, "noocytes," and after he injects them into his own bloodstream, they evolve though "self-directed development."[55] Eventually they begin communicating with Ulam while transforming all his cells into copies of themselves. Ulam collapses into a puddle of goo, a fluid mass of noocytes freed from the formal boundaries of the body. Other people become "infected" with the noocytes, a plague of "transformation and dissolution takes place within a week of infection" (160), and soon the human population of

North America dissolves into continent-spanning flows of intelligent cells, "a landscape of biological nightmare" (229).

The noocytes exist both autonomously and in manifold cooperative assemblages, from small cell clusters to enormous mountains of tissue, constantly swapping constituents, sharing information with each other through the "blood music" of chemical signals: a rhizomatics where new structures and properties emerge out of the reshufflings of biomass. But the cellular horizon of life also breaks down when the noocytes become aware of the "life" of their own internal components: "Their cytoplasm seems to have a will of its own. A kind of subconscious life, counter to the rationality they've acquired so recently. They hear the chemical 'noise' of molecules fitting and unfitting inside" (97). The molecular structures inside their own single-celled bodies evince "a kind of subconscious life," a "flow and electric sensation of pure life" beyond the ordered mastery of the organism (262), beyond the mastery of the cell. Disintegrating into themselves, "sinking deeper into the noocyte world, discovering layers and layers of universes within" (301), the noocytes learn that not only can living entities dispense with the organism and the cell, but they can even dispense with the molecule: life, it turns out, can self-organize well below the crudities of the "chemical" realm: "Information can be stored even more compactly than in molecular memory. It can be stored in the structure of space-time. What is matter, after all, but a standing-wave of information in the vacuum?" (330). Therefore, to remove environmental barriers on their reproduction and expanding intelligence, the noocytes disintegrate themselves into the quantum probability waves of materiality, destratifying the universe entirely and reorganizing it to their own specifications.

Blood Music thus disintegrates the organism into its cells, the cells into their molecules, and finally the molecules into the quantum structure of their own atoms. Like nesting dolls, the scales at which life manifests in this novel diminish from the macro to the micro to the nano to the pico, finding greater space for life, for the endless reproduction and expansion of life's horizon in the smaller and smaller worlds within. The narrative makes good on Richard Feynman's promise—the foundational promise of nanotechnology—that "there's plenty of room at the bottom." Which turns out to be the same promise made by splatter fiction: "There are no limits."[56]

Life in *Blood Music* disintegrates to such an extent that it is no longer anywhere, because it is everywhere, and the universe as a contingent organization of space-time is now restructured by the physical transcendence of intelligent life beyond matter, remade in its very possibility-structures to accommodate the ever-expanding, ever-growing desires of machinic entities for a space in which to live. The novel concludes that "there is no ultimate reality" of the universe (332), except the reality literally constructed by the desires of life.[57] *Blood Music* exposes the philosophical implications of nanosplatter as a postbiological discourse: through serial disintegration (no longer, we see, suggesting the collapse or implosion of space, but rather its unlimited expansion), life manifests at lower and lower levels until there is nothing left but *more life*. Life, discontent to inhabit a known universe, instead changes the universe—changes "life itself"—in order to spread and grow. Life now marks a surplus of possibility as the abstracted remainder of disintegration; it is the stain left behind when even the universe is destratified. This is the end of things seen through the furthest speculations of nanofiction: not death, but life . . . life deterritorialized from every possible assemblage, expanding beyond the organism, beyond the molecule, beyond the confines of the real.

THE SURPLUS LIFE OF THE NANOSPLATTERED BODY

The fictive experiment of nanosplatter echoes the philosophical experimentation of Deleuze and Guattari in shattering molar aggregates into multiplicities and laying open the "domain of nondifference between the microphysical and the biological, there being as many living beings in the machine as there are machines in the living."[58] Discovering this "domain of nondifference" beyond the blasted organism, nanofiction sees the postbiological not as the end of life but as the condition of "surplus life," the "excessively biological," or the overflow of life beyond itself. In other words, the gap made in the symbolic unity of the living at the scene of disintegration is not a void but a fecundity that gives birth to a postbiological riot.

As we have seen in *Blood Music*, at every level of disintegration, at every approach to the "final" level in which life can manifest, there is always a deeper internal depth tempting the noocytes with a glimpse of "more life." Instead of reaching an exhaustion or a limit of life—instead of reaching a "minimalism of the organism"—at every stratum of the noocytes' organic archaeology we see a blossoming of the forms and numbers of entities populating the universe: novel organismic assemblages, cells, molecules, quantum wave-packets, and even a Chardinian "Noosphere" are thrown off as enlivened creatures during the course of the noocytes' odyssey. The postbiological would thus describe the profligate production of alternative modes of living—machinic life, technogenic life, nonorganic life, nanotech life, quantum life, galactic life, and so forth—as a direct consequence of the singularity of "life itself" giving way to its outside.

Offering surplus in the place of limit, multiplicity in the place of singularity, the postbiological presents itself as an epistemic alternative to the restrictive binary logic of 1 and 0, presence and absence, living and nonliving. This is suggested quite graphically in the final chapter of *Bloom*, when the mission captain decides to unpack himself and join the expanses of technogenic entities. This splattering significantly takes place within a textual abyss formed by the incompletion of the chapter's aphoristic title: "Twenty-five: If You Can't Beat Them . . ." (298; ellipses in original). That is to say, the representation of the captain's unpacking that follows these ellipses is not identical to the presumed-missing phrase, "Join Them"; it does not "fill in" the gap just to reinforce its emptiness, but inasmuch as the novelistic depiction of the captain's disintegration suggests "joining them," it works as a narrative alternative to any (humanistic) desire for completion potentially produced by a perception of lack in the chapter title. Instead of completion, the novel offers a "joining" in multiplication, even at the formal level joining its scene of disintegration (its scene of "joining them") to the incompletion of the chapter title as an excess-production of joinings. The captain's unpacking is a fictional representation of postvital abundance and machinic connectivities, generated in and joined to the place of absence, a multiplicity that expands beyond the site of pure negativity. Or, as the captain explains, "Better than death, I think" (302). Which is, I should stress, not a logocentric solution to a panic-inducing absence, but rather an effort to grow beyond the very logocentrism that would make "life itself" a unity of full presence and death the only alternative.[59]

The postbiological escapes death, then, by discovering more (than) life in the place of absence, and thus finds itself wanting nothing. As the Mycosystem exclaims: "We do not *want*. . . . We have grown beyond it" (297). They have—they are—plenty. They are in excess as such: "grown beyond it." Certainly they do not lack "life" because they have discovered that life, always in excess of itself, can be exploded to accommodate "technogenic life." But this is not to say that they are not desiring; indeed, they are quintessential examples of the *desiring-machines* that, as Deleuze and Guattari have written, become visible in the disintegration of those molar organizations (like the organism, or market economies) that channel desire through lack: "Once the structural unity of the machine has been undone, once the personal and specific unity of the living has been laid to rest, a direct link is perceived between the machine and desire, the machine passes to the heart of desire, the machine is desiring and desire, machined."[60] Desiring-machines do not lack anything because their "object of desire is another machine connected to it," suggesting that the molecularity of desire is not lack but production, the desiring-production of machines connected to machines.[61] Hence the entities that make up Mycosystem, at the moment of expressing their passage through "want," also suggest their desiring-production and preservation of the "constituent and support complexes" that they connect to (297), that they instantiate, and that generate their technogenic lives. They manifest, then, a kind of machinic desiring-production "beyond" desire-as-lack, across the disintegrated and deterritorialized limits of humanistic desire (phallogocentrism). As such, the Mycosystem is radically outside any desire comprehensible to "the human": "It is no patronization to say, we have concerns you will not comprehend" (297).

In registering the existence of this postbiological desiring-realm in excess of biological limits, even—or, I should say, especially—if it remains "incomprehensible," nanofiction encodes a trace of absolute otherness, of machinic alterity, into its more-or-less traditional framework of novelistic conventions and humanist perspectives. It is at this moment of incomprehension, in which otherness is recognized in its otherness but remains uncolonized, in which the beyond of biologistic discourse emerges not from without but from within the very space of biologism, that the nanofiction narrative reveals itself as a "literary assemblage" of discourses: not a monadic "book" governed by a single humanist perspective, but an "information multiplicity" (to use John Johnston's term) of multiple signifying and asignifying (or incomprehensible) regimes inventoried within the same textual system.[62] The nanofiction narrative therefore performs the postbiological inhabitation of the biological at the discursive level, as an inner excess, a trace of "beyondness" announcing itself within the linear narrative.

The text itself, encoding the discourse of the postbiological other as simultaneously within *and* beyond the comprehension of its own discourse, thus mirrors its figuration of postbiological embodiment emerging within biological embodiment—by which I mean both the disintegration of the organism and the surplus continually discovered in life's place. At the discursive and the figurative level, the postbiological is the beyondness already inside the biological, requiring only an act of disintegration to unbind it. Or as *Blood Music* puts it: "They're inside, part of us by now. They are us. Where can we escape?" (138). The postbiological, then, is less the "afterlife of life" than the "extralife of life," the "otherlife of life"—the otherness already immanent to biologism that, in the nanofiction text, inscribes itself within the place of biology while retaining its otherness. It therefore appears that what nanofiction expresses—even in its most paranoid depictions of nanotechno-logical horrors—is a fundamental openness to the other already "inside, part of us by now." I would even suggest that this openness to the other inside begins to perform as something very much like *love*.

Nanofiction as love, then . . . love for the nanomachines. A strange idea, perhaps, especially considering that this love would express itself for entities that not only do not exist—or at least do not yet exist—but also whose fictive emergence regularly depends on violently dismantling the human body. But we can begin to see that love is the issue, or what is being issued, inasmuch as the human becomes imagined as carrying the other inside itself and gives of itself through disintegration, literally opening up to those entirely ficti-tious others still awaiting the future for their realization. As *Crawlers* puts it: "It's funny how technology takes on a life of its own. . . . It's like we surren-der some of our own life to technology" (323). *Crawlers* suggests that life is not a thing that we might "possess," that we could sever from ourselves and give to technology as the fulfillment of its desire—life turns out not to be an object—because it is fundamentally *divisible* and *multiple*. If we "surrender *some* of our own life to technology," then there would be more life in us than ourselves; there would be an excess of life that can be fragmented, multi-plied, broken into parts, in a way that does not subtract from the whole. This shattering of life into parts (partial-objects) and "surrendering some of it" to the nanomachine, then, is not a giving only to be taken away (like humanis-tic love, which, as Lacan has shown, "gives what it doesn't have"), but rather a giving through which the nanomachine is able to manifest a life "of its own." It is a giving of postbiological abundance in the place of biological scarcity, a giving of the multiple and the surplus in the place of the singular and the exclusive: an issuing of "lives" in place of "life."

The splatter-disintegration of the body thus literalizes a love that does not simply give *to* the other but gives *of* the other as the alterity in itself, gives *up* the other as an excess that both is, and is not, itself. So it is less a giving than an opening—an *unpacking*—that exposes itself to the encounter with difference, excavating the others within, releasing a cornucopia of others in its discovery that "you are utterly free. . . . Free to Unpack, if you choose" (*Bloom*, 297). It is—for they lived happily, after all—a story of *posthuman love*. While it may pose as a horror story from the perspective of biologistic humanism (and many nanofiction narratives ostensibly depict it as such), nanovision sees the situation otherwise. As Greg Bear has reflected, when it comes to the scene of disintegration, "What at first seems an unmitigated horror is in fact much more, if we could only take off the blinders of our mortal individuality."[63] The splatter-technology of nanovision frees love from its infamous blindness, from its human remains. The deconstruction of the organism necessarily opens itself to those machinic others yet to come, and those machinic others already inside of it.[64]

Nanofiction therefore makes a spectacular set piece of Derrida's assertion that "deconstruction . . . is an openness to the other."[65] And of course, "Deconstruction . . . never proceeds without love."[66] In its disintegration of the biological, its deterritorialization of vital matters from the autopoietic closure (or narcissism) of the organism, it inscribes an anticipatory trace of the technogenic other already "inside us, part of us by now." It suggests, indeed, that "they are us." Nanofiction is thus a site of our becoming-alien to ourselves—a site of xenomorphosis, or what Ann Weinstone has called the "avatar body." An ethical space of posthuman devotion where self and other intersect, an "undecidable zone of relationality," the avatar body is characterized by "a nonheroic, unconditional hospitality, an indiscriminate, noncharismatic, and deliberate offering that can never resolve into static sameness or absolute difference."[67] The avatar body of nanofiction is made "semiotic flesh" at the interface of the text (a polyglossia of human and nonhuman discourses encoded in the same literary assemblage) and the imaginary splattered body of disintegration (the organism opened to its molecular others): textual and conceptual, the avatar body emerges as a "thoroughfare of relations" where human and nonhuman identities pass through each other (160), constantly decomposed into globs of self-other. The nanosplattered avatar body is a vehicle through which postbiological otherness takes up residence and finds a room of its own (plenty of room!) inside the human, which is therefore no longer simply the human.

For the avatar body draws the reader into its ethical zone of relationality (as one of its many inhabitants) by asking us to recognize our *own* bodies as intersections between the human subject and the field of self-motivating matter that inhabits our cells but overflows our selves. By flaying the body open to awareness of the fraction of the machinic phylum that traverses the human subject but continually exceeds it, nanofiction *itself* works as a splatter-technology. That is, in its being read (red?), it acts as a literary-machine engineering possibilities of embodiment in which multiplicities of lives can exist together beneath the same tattered skin. Even now.

The "10,000 different nanomachines at work inside our bodies," sent via semiotic post toward their becoming-entity, embody the machinic otherness already inside us. They are the postbiological abiding within the biological. And nanofiction, in modeling the human body as a place of encounter between the human self and the machinic others sharing the same space, thereby enables us to imagine containing more life than ourselves, to actually make for ourselves a "body without organs"—a materialization of the avatar body—which is "not an empty body stripped of organs, but a body upon which that which serves as organs . . . is distributed according to crowd phenomena, in Brownian motion, in the form of molecular multiplicities."[68] In *Prey*, Jack Forman's encounter with the nanoswarms enables him to recognize the infrahuman "swarm intelligence" that he already comprises, and he begins to think himself a body without organs, a body without a governor, a body without *control*:

> The control of our behavior is not located in our brains. It's all over our bodies. . . . "Swarm intelligence" rules human beings . . . and rarely comes to consciousness. . . . And for that matter, a lot of sophisticated brain processing occurs beneath awareness. . . . The whole structure of consciousness, and the human sense of self-control and purposefulness, is a user illusion. We don't have conscious control over ourselves at all. We just think we do. (260–61)

Identifying the realm of self-motivating matter beneath our "structure of consciousness," nanofiction translates this material flux of the body—the body which is "actually a giant swarm . . . a swarm of swarms" (260)—and accommodates its extrahuman autogenesis: "And for all we knew, this damned swarm had some sort of rudimentary sense of itself as an entity" (261). It stages the phenomenological impact of this swarm that is simultaneously within and without the human, this multiplicity of desiring-entities cohabitating the now-postbiological body alongside the human "I." "Just because human beings went around thinking of themselves as 'I' didn't mean it was true" (261); in reality, we contain multitudes. And nanofiction writes them as they "come to consciousness," announcing to us that we are not alone inside ourselves. So while Jack Forman fights the symbiotic aspirations of nanoswarms in order to preserve the future (as his surname indicates) "for man," it seems this battle has already been lost.

Because if these novels work to perform the *postbiology of the human in its experiential encounter with the text*, they enact the epistemic preconditioning necessary for the nanomachine to be recognized as "technogenic life" in the future. Similar to the way that some nanobiologists discern nascent traces of the future, signs of machinic life, contained within current successes of biomolecular engineering—thereby tautologically constructing the conditions of possibility for "living technologies" to eventually emerge—nanofiction too projects a technogenic future from within the zone of self-alienation, or love, that it bodies forth. It proceeds according to the "postness" of the postbiological and puts an anticipatory trace of the technogenic inside our bodies. The biological is thus stamped with its future as the surplus otherness unfolding in itself.

For if the postbiological is a posting, a sending, a giving of its destiny toward a recipient future beyond the biological, then by "addressing" the future it seems already to entail that future (I am, of course, referring to Derrida's analysis of the postal logic of destining).[69] Its "destination" appears inscribed on the symbolic surface of the present as an effect of narrative speculation (since narrative always entails some sense of destining . . . "the end" has already been written). As if our present—even the "gift" of (some of) our biology—has already been opened to the future by the speculations of our own narratives. As if, like so many visionary accounts of nanotechnology have claimed, the nanofuture has *already* become inevitable. Charles Ostman writes of this destining: "What appears to be the future for many, is already the past for some, who are now unfolding it into the present."[70] Or, as John Robert Marlow puts it in his authorial postscript to *Nano*: "It *cannot* be prevented" (369).

But at the same time, it is the posting of this future that opens it to the possibility of otherness—even, indeed, to the possibility of this letter *not* reaching its destination—and in negotiating a future destined to nothing other than its own possibility of becoming *different*, the specularity of the post opens to the otherness immanent in itself. For by posting the postbiological, nailing it up for us to see (if only from the corners of our eyes), nanofiction also makes us aware of the limitations of a presentism that blinds us to *possible* realms of otherness that may yet come, or that may yet already exist, at the exobiological extremes of cosmic distance or the nanobiological extremes of terrestrial interiority. Do nanobacteria have a future? Do nanomachines? Along with speculative nanoscience, the experiments of nanofiction work to make these futures at least possible, even now. As Brian Attebery has written, nanofiction "convey[s] messages from the quantum world, which is also the world in which we have been living all along without knowing it."[71]

Or to think of it another way: even as science (fiction) speculates and splatters the edges of the present, we become increasingly able to recognize the features of the nanoscale world through and beyond our own biological lives, and thus we step up to the "dance of agency" with the machinic phylum, both dancing partners made equal in love, opening to each other.

After all, we engineer the future only by discovering the boundaries of our own time: our limits of fabrication, our failures of imagination, our terminal singularities. By finding them circumscribed as limits, held together only by surface tensions, we encounter the trace or the touch of the alien outside, which is therefore no longer really outside or quite so impossible, even if it hardly matters anymore what it really means to be real. "We're standing on the threshold," writes Linda Nagata in her novel *Limit of Vision* (2001), where the "on-ramp to nanotech" emerges through "the first artificial life-form, a symbiotic species affectionately known as LOVs—an acronym for Limit of Vision, because in size LOVs are just at the boundary of what the human eye can easily see."[72] A very small mote in the eye—a figment of the onrushing age of nanotechnology in all its "artificial" vitality, all its technical fictivity— appears right at the edge of our blindness, along that very boundary between our limited selves and otherwise, tantalizing with promises of the beyond, promises of "LOV" in all its senses, with more on the way.

It's coming. Or rather . . . it's here.

NOTES

INTRODUCTION

1 Vinge, "Technological Singularity," 88–95, 89, 90; hereafter cited in the text as "TS." Originally presented at the 1993 Vision-21 Symposium (sponsored by NASA Lewis Research Center and the Ohio Aerospace Institute), this article reworks and updates Vinge's earlier thoughts on Singularity in his *Omni* article "First Word" (1983).

2 Hawking, *A Brief History of Time*, 122.

3 Vinge follows the mathematician John von Neumann in positing such a radically estranging historical "singularity." As reported by Stanislaw Ulam in his tribute "John von Neumann, 1903–1957": "One conversation [with von Neumann] centered on the ever-accelerating progress of technology and changes in the mode of human life, which gives the appearance of approaching some essential singularity in the history of the race beyond which human affairs, as we know them, could not continue" (5).

4 Vinge's fictional explorations of the Singularity include "'Bookworm, Run!'" (1966), *The Peace War* (1984), *Marooned in Realtime* (1986), *A Fire upon the Deep* (1992), and several other stories.

5 Vinge, *Marooned in Realtime*, 176 (ellipses in original), 265, 177.

6 Broderick, *The Spike*, 83, 15, 83.

7 Kurzweil, *The Singularity Is Near*, 368.

8 Kurzweil, "After the Singularity," 141–51, 150. Kurzweil's books on advanced machine intelligence and other developing sciences have consistently imagined a technological Singularity in the very near future; see *The Age of Intelligent Machines* (1990), *The Age of Spiritual Machines* (1999), and, with Terry Grossman, *Fantastic Voyage* (2004).

9 See Moravec, *Mind Children*, and *Robot*. On the various communities (or cults) of futurists who embrace Moravec's concept of uploading—the technophilic faithful who often arrange their lifestyles, assets, and social networks around the notion of silicon transcendence—see Dery, *Escape Velocity*, 299–319; and Doyle, *Wetwares*, 121–42.

10 Moravec, "Singularity Equation Correction," 137–40, 140.

11 Benford, "Comment by Gregory Benford," in R. Hanson, "A Critical Discussion of Vinge's Singularity Concept."

12 More, in Kurzweil and More, "Max More and Ray Kurzweil on the Singularity," 154–70, 155–56.

13 Hayles, *How We Became Posthuman*, 3.

14 Broderick, *The Spike*, 56.

15 Vinge, afterword to *Marooned in Realtime*, 271.

16 See G. Milburn, *The Feynman Processor*; and Drexler, "Machine-Phase Nano-technology," 74–75.

17 Broderick, *The Spike*, 128. Nanotechnology's privileged role in speculations on the Singularity derives in part from its potential as an enabling technology that could lay the groundwork for even more profound technical changes, including the development of true superintelligence. As the Singularity Institute for Artificial Intelligence explains:

> From the perspective of the Singularity, nanotech is the prototypical example of *rapid infrastructure*. Nanotech can self-replicate; it moves physical manipulation into a new timescale — microseconds instead of seconds — and thus acts as a very rapid base for the acquisition of further ultratechnologies; and it turns matter into information, allowing in a sense the "reprogramming of reality." . . . The point of discussing nanotechnology is to show that technologies accessible outside the human regime are significantly more powerful than those we are accustomed to dealing with; they can bypass existing human infrastructure; operate on timescales far quicker than human hands; and provide a substrate for thought tremendously faster than both existing human brains *and* existing computer technology. ("Nanotechnology," n.p.)

Science fiction has also frequently depicted nanotechnology as fueling the Singularity; for example, see Rucker, *Postsingular*; and Stross, *Singularity Sky*.

18 On the multidisciplinary nature of nanotechnology, see Schummer, "Multi-disciplinarity, Interdisciplinarity, and Patterns of Research Collaboration in Nanoscience and Nanotechnology," and "Interdisciplinary Issues in Nanoscale Research." Schummer argues that nano's multidisciplinarity masks a deeper inability to achieve adequate interdisciplinarity, owing largely to the competing research programs and radically incommensurable paradigms that divide the various fields of nanoscience. On the formation of new disciplinary identities in nanoscience that simultaneously accommodate and expand the boundaries of traditional research fields, see Kurath and Maasen, "Toxicology as a Nano-science?"

19 See Taniguchi, "On the Basic Concept of 'Nano-technology'"; and Feynman, "There's Plenty of Room at the Bottom." The question of nanotechnology's origins will be taken up in chapter 1.

20 "Exploratory engineering" and "theoretical applied science" are terms coined by Drexler to describe anticipatory scientific research in which technological reality has yet to catch up; see Drexler, *Nanosystems*, 489–506. "Exploratory engineering" presents itself as a technical discourse that narrativizes scientific possibility to guide the future development of nanotechnology. As we will see, narrative plays a central role in nanoscience generally, from exploratory engineering to the digital transcoding of molecular structures by scanning probe microscopes, whose translated raster patterns are themselves forms of narrative; see Pressman, "Nano Narrative."

Exploratory engineering investigates future technologies by theorizing them, narrating their design, applications, and consequences while mathematically, computationally, or mechanically simulating them to assess viability. The simulation aspect of exploratory engineering occurs sometimes through material models, but more often on powerful computers. As a promotional report from the U.S. National Science and Technology Council put it, nanoscientists "rely heavily on extremely powerful computers to model and simulate nanoscale structures and phenomena. This helps them use their theories better and to make sense of what they see through their [scanning probe] microscopes. Simulation and modeling also help them evaluate the infinite number of possible nanostructures they could in principle try to build. Simulations can help steer researchers toward fruitful directions and away from dead ends" (*Nanotechnology: Shaping the World*, 6). On the role of simulations in nanoscience, see A. Johnson, "The Shape of Molecules to Come"; Lenhard, "Nanoscience and the Janus-Faced Character of Simulations"; and Winsberg, "Handshaking Your Way to the Top." Nanotechnology's heavy reliance on narrative, simulation, and exploratory engineering has consequences for the way we understand its relationship to the real, as we will see in chapter 1.

21 W. Patrick McCray shows that futuristic utopian visions have characterized the entire history of nanotechnology and, moreover, have played a fundamental role in the development of funding legislation, especially the monumental U.S. National Nanotechnology Initiative; see McCray, "Will Small Be Beautiful?"

22 See National Science and Technology Council, *National Nanotechnology Initiative*; and Roco and Bainbridge, *Converging Technologies for Improving Human Performance*.

23 Smalley, "Nanotechnology: Prepared Written Statement," 55.

24 See Ratner and Ratner, *Nanotechnology: A Gentle Introduction to the Next Big Idea*; and Atkinson, *Nanocosm: Nanotechnology and the Big Changes Coming from the Inconceivably Small*.

25 See Berube, *Nano-Hype*.

26 Gubrud, interview by Olson.

27 On the technological determinism of nano, see Mody, "Small, but Determined." On catastrophic risk perception as a consequence of this nanodeterminism, see Marshall, "Future Present."

28 Gubrud, interview by Olson.

29 Marlow, "The Sound of Inevitability."

30 Winner, *Autonomous Technology*.

31 Kurzweil, *The Singularity Is Near*, 486–87.

32 On the technological reductionism characterizing "nanotechnology" as the site of convergence for numerous scientific and engineering disciplines, see Schmidt, "Unbounded Technologies." To the extent that this convergence toward a nanotechnological worldview of systemic complexity can itself be accounted for by a unified systems theory perspective, see Khushf, "A Hierarchical Architecture for Nano-scale Science and Technology."

33 Elliott, preface to *Nanodreams*, 2.

34 Nishi, quoted in Bruno, "The Next Big Thing Is Really Tiny," 48.

35 This deconstructive projection of blindness as itself the occasion for spectacular insights has been famously theorized by Paul de Man in his *Blindness and Insight*.

36 My use of the concept of "ways of seeing" follows Berger, *Ways of Seeing*, in the sense of technical conventions and epistemological conditions that enable vision to take place within a certain cultural environment. We might also compare the "paradigms" described in T. Kuhn, *The Structure of Scientific Revolutions*; or the "thought collectives" described in Fleck, *Genesis and Development of a Scientific Fact*; or the "scopic regimes" described in Metz, *The Imaginary Signifier*, and in Jay, *Downcast Eyes*.

37 Lyotard, *The Postmodern Condition*, xxiv.

38 Jacques Derrida frequently critiqued the apocalyptic tone in supposedly anti-humanist theoretical discourse, seeing therein a reinscription of the very humanism that such discourse would pretend to escape. See Derrida, "The Ends of Man," and "Of an Apocalyptic Tone Recently Adopted in Philosophy." Also see his "No Apocalypse, Not Now."

39 Dery, *Escape Velocity*, 8; Doctorow, "The Rapture of the Geeks." See also Alexander, *Rapture*. Alexander documents the quasi-theological aspects of various transhumanist movements, including their widespread faith in the rapturous arrival of the Singularity. On transcendent longing across the fields of contemporary technoscience, see Stafford, "Leveling the New Old Transcendence." On the rapturous and apocalyptic undertones of nanotechnology discourse itself, see Schummer, "Nano-Erlösung oder Nano-Armageddon?"

40 On becoming and evolution as movement processes that defy static presentism—and thus defy metaphysical regimes that would close only with a sublime rapture—see Deleuze and Guattari, *A Thousand Plateaus*; and Massumi, *Parables for the Virtual*.

41 More, in Kurzweil and More, "Max More and Ray Kurzweil on the Singularity," 157. For a related discussion on the need to distinguish Singularity theory from theology despite surface analogies, see Kurzweil, *The Singularity Is Near*, 369–90.

42 On rapturous humanism and the persistence of the "liberal humanist subject" within some forms of posthuman thought, see Hayles, *How We Became Post-human*. Hayles exposes posthumanism's blindness to its own internal human-ist biases while simultaneously showing how it creates moments of innovation and novelty, suggesting a complex seriation of the posthuman with the human. By pursuing some of the fractures created in this seriation, Hayles offers alter-native possibilities for posthuman thought in which the body remains. Also see Badmington, *Alien Chic*, 34–63, 109–99. Badmington argues that the human always inhabits the posthuman (even or especially) in its most apoca-lyptic breaks, and therefore "posthumanism" is always a "working through" of humanism toward becoming-posthuman, rather than a separate or alternate site of resistance.

43 McCarthy, *Bloom*, 72.

44 Templesmith, *Singularity* 7, 10–12.

45 Badmington writes, "Humanism is there and not quite there. It comes and goes, it flickers, it drifts, and it is precisely this wandering that I want to call the possibility of posthumanism. . . . The 'post' does not mark an end, a break, or novelty; it identifies, rather, a patient reckoning with—a working through of—what follows the prefix." He therefore concludes, "Posthumanism, as I see it, is the acknowledgement and activation of the trace of the inhuman within the human" (*Alien Chic*, 145, 157). On posthumanism as an ethical opening to "the other-than-human [that] resides at the very core of the human itself," see C. Wolfe, *Animal Rites*, 17.

1. THE AGE OF POSTHUMAN ENGINEERING

1 Drexler, preface to Drexler, Peterson, and Pergamit, *Unbounding the Future*, 10.

2 Roco, "Nanotechnology's Future," 39. Roco is one of the key architects behind the U.S. National Nanotechnology Initiative.

3 Mirkin, "Tweezers for the Nanotool Kit," 2095.

4 Service, "AFMS Wield Parts for Nanoconstruction," 1620.

5 Service, "Borrowing from Biology to Power the Petite," 27.

6 Gimzewski and Joachim, "Nanoscale Science of Single Molecules Using Local Probes," 1683.

7 Smalley, "Nanotech Growth," 37.

8 Peterson, "Nanotechnology: Evolution of the Concept," 186. Indicative of nano-writing's teleological tendencies, Peterson's article absorbs the entire history of atomic theory, from Democritus to the present, to suggest the unavoidable rise of nanotechnology and our progression toward the nanofuture.

9 Kurzweil, *The Singularity Is Near*, 407.

10 Moravec, review of *Engines of Creation*, 76–77.

11 Cramer, "Nanotechnology: The Coming Storm," 8.

12　In *Nanosystems*, Drexler defends the usage of present-tense narration for nano-writing:

> In ordinary discourse, "will be" suggests a prediction, while "would be" suggests a conditional prediction. Using these future-tense expressions is inappropriate when discussing the time-independent possibilities inherent in physical law. . . . The present tense is more serviceable: One can say that as-yet unrealized spacecraft trajectories to Pluto "are of two kinds, direct and gravity assisted," and then analyze their properties without distraction. Similarly, one can say that as-yet unrealized nanomachines of dimondoid structure "are typically stiffer and more stable than folded proteins." Much of the discussion in this volume is cast in this timeless present tense; this is not intended to imply that devices like those described . . . presently exist. (xix)

Nanowriting employs literary techniques common to speculative science writing in general. See Myers, "Scientific Speculation and Literary Style in a Molecular Genetics Article," on the linguistic peculiarities of scientific specu-lation that work to legitimate such claims. On the ability of speculative rhe-toric to orient technical development, see van Lente and Rip, "The Rise of Membrane Technology"; and Sunder Rajan, *Biocapital*, 107–181. Nanowriting, however, goes beyond most scientific speculation in that its uses of the future tense and its visions of tomorrow are totalizing, bringing the long-range future completely into the textual present. It depicts not just one possible technology of the future but rather the future itself as an effect of this "inevitable" tech-nology—which is one reason, as we will see, why nanotechnology has so often been characterized not as "speculative science" but as "fictional science." For more on the features of nanowriting and its modes of technical forecasting, see Faber, "Popularizing Nanoscience"; and Schummer, "Reading Nano."

13　Crandall, preface to *Nanotechnology: Molecular Speculations on Global Abun-dance*, ix.

14　Drexler, "Molecular Engineering," 5278.

15　Eigler, quoted in Service, "Atom-Scale Research Gets Real," 1526.

16　D. Jones, "Technical Boundless Optimism," 835, 837.

17　Stix, "Trends in Nanotechnology," 97.

18　Stix, "Little Big Science," 37.

19　Whitesides, "Nanotechnology: Art of the Possible," 85.

20　Whitesides, "The New Biochemphysicist," 24.

21　Block, "What Is Nanotechnology?"

22　Many early critiques of nanotech's "science fictionality" are described in Regis's lively history *Nano: The Emerging Science of Nanotechnology*.

23　Recent scholarship in science studies has increasingly noticed a construc-tive bleed between nanotech and science fiction. See Miksanek, "Microscopic Doctors and Molecular Black Bags"; C. Milburn, "Nanotechnology in the

Age of Posthuman Engineering"; Gimzewski and Vesna, "The Nanomeme Syndrome"; Hayles, *Nanoculture*; López, "Bridging the Gaps"; Hessenbruch, "Nanotechnology and the Negotiation of Novelty"; Schummer, "Societal and Ethical Implications of Nanotechnology"; Toumey, "Narratives for Nanotech"; Coenen, "Nanofuturismus"; Paschen et al., *Nanotechnologie*, 257–74; Hessenbruch, "Beyond Truth"; Nerlich, "From Nautilus to Nanobo(a)ts"; C. Milburn, "Nanowarriors"; Lösch, "Anticipating the Futures"; Thurs, "Tiny Tech"; and Bowman, Hodge, and Binks, "Are We Really the Prey?" On the potential for further collaboration between nanoscience and literature, see Avery, "Nanoscience and Literature."

24 Suvin, *Metamorphoses of Science Fiction*, 64.

25 Ibid., 75.

26 The fabulation of worlds and zones of radical otherness characterizes the interplay between science fiction and postmodernist writing; see McHale, *Postmodernist Fiction*, 59–72.

27 Baudrillard, *Symbolic Exchange and Death*, 50–86. See also Baudrillard, "The Precession of Simulacra."

28 Baudrillard, "Simulacra and Science Fiction," 121–22.

29 Ibid., 124, 125, 126. For related analysis of this collapse of futuristic fiction into present reality as a postmodern symptom, see Csicsery-Ronay, "Futuristic Flu."

30 For various other examples of constitutive feedback between science and science fiction, see Balsamo, *Technologies of the Gendered Body*; B. Clarke, *Energy Forms*; B. Clarke, *Posthuman Metamorphosis*; Dick, *The Biological Universe*; Disch, *The Dreams Our Stuff Is Made Of*; Doyle, *Wetwares*; Franklin, *War Stars*; Gray, "There Will Be War!"; Hamilton, "Traces of the Future"; Haraway, *Primate Visions*; Haraway, *Simians, Cyborgs, and Women*; Haraway, *Modest_Witness*; Hayles, *How We Became Posthuman*; Hayles, *My Mother Was a Computer*; S. Johnston, *Holographic Visions*; Kilgore, *Astrofuturism*; Markley, *Dying Planet*; Mellor, "Between Fact and Fiction"; Nahin, *Time Machines*; Otis, *Membranes*; Pendle, *Strange Angel*; Penley, *NASA/Trek*; Squier, *Liminal Lives*; Thacker, "The Science Fiction of Technoscience"; Turney, *Frankenstein's Footsteps*; and Willis, *Mesmerists, Monsters, and Machines*. On the role of futurology and speculative visions in shaping technoscience, see Nik Brown, Rappert, and Webster, *Contested Futures*; and Struken, Thomas, and Ball-Rokeach, *Technological Visions*.

31 Haraway, "A Cyborg Manifesto," 149.

32 Bukatman, *Terminal Identity*, 22.

33 Hayles, *How We Became Posthuman*, 3.

34 Merkle, "It's a Small, Small, Small, Small World," 26.

35 Montemagno, "Nanomachines," 1.

36 These explicitly science-fictional images and a general faith in the imminent nanofuture can be found in (but are certainly not limited to) Drexler, "Molecular Engineering"; Drexler, "Molecular Manufacturing as a Path to Space"; Merkle,

"Nanotechnology and Medicine"; Freitas, *Nanomedicine*; Heckl, "Molecular Self-Assembly and Nanomanipulation"; Freitas and Merkle, *Kinematic Self-Replicating Machines*; Colbert and Smalley, "Fullerene Nanotubes for Molecular Electronics"; Yakobson and Smalley, "Fullerene Nanotubes"; Hameroff, *Ultimate Computing*; Sargent, *The Dance of Molecules*; Ho, Fung, and Montemagno, "Engineering Novel Diagnostic Modalities and Implantable Cytomimetic Nanomaterials for Next-Generation Medicine"; and Hall, "Utility Fog."

37 This argument is ubiquitous in nanowriting. Some stronger instances include Drexler, *Engines of Creation*, 5–11; Drexler, "Molecular Engineering," 5575–76; Smalley, "Nanotechnology: Prepared Written Statement"; Merkle, "Molecular Manufacturing"; Merkle, "Self-Replicating Systems and Molecular Manufacturing"; Service, "Borrowing from Biology to Power the Petite," 27; Montemagno, "Nanomachines"; Goodsell, *Bionanotechnology*; and R. Jones, *Soft Machines*.

38 Roco, "International Strategy for Nanotechnology Research and Development," 356, fig. 2. Roco suggests that nanotech, now leaving its science fiction baggage behind, will progress through four "evolutionary stages": research on the properties of passive nanostructures; development of active nanostructures; cultivation of nanostructure systems; and finally, "after 2015–2020, the field will expand to include molecular nanosystems—heterogeneous networks in which molecules and supramolecular structures serve as distinct devices. The proteins inside cells work this way, but . . . these molecular nanosystems will be able to operate in a far wider range of environments and should be much faster. Computers and robots could be reduced to extraordinarily small sizes. Medical applications might be as ambitious as new types of genetic therapies and antiaging treatments. New interfaces linking people directly to electronics could change telecommunications" ("Nanotechnology's Future," 39). It would seem, then, that nanotech leaves its science fiction baggage behind only to pick it up again in the cyborg future.

39 Mody, "How Probe Microscopists Became Nanotechnologists."

40 Binnig et al., "Tunneling through a Controllable Vacuum Gap"; Binnig et al., "7 X 7 Reconstruction on Si(111) Resolved in Real Space."

41 DeGrado, Regan, and Ho, "The Design of a Four-Helix Bundle Protein"; Regan and DeGrado, "Characterization of a Helical Protein Designed from First Principles."

42 Foster, Frommer, and Arnett, "Molecular Manipulation Using a Tunnelling Microscope."

43 Eigler and Schweizer, "Positioning Single Atoms with a Scanning Tunnelling Microscope."

44 Yakobson and Smalley, "Fullerene Nanotubes"; Dai, Franklin, and Han, "Exploiting the Properties of Carbon Nanotubes for Nanolithography"; Kim and Lieber, "Nanotube Nanotweezers."

45 Davis, "Synthetic Molecular Motors"; Cuberes, Schlittler, and Gimzewski,

"Room-Temperature Repositioning of Individual C_{60} Molecules at Cu Steps";
J. Knight, "The Engine of Creation"; Asfari and Vicens, "Molecular Machines."

46 Lee and Ho, "Single-Bond Formation and Characterization with a Scanning
Tunneling Microscope."

47 Gorman, Groves, and Shrager, "Societal Dimensions of Nanotechnology as a
Trading Zone." Nanotechnology in this account is a multidisciplinary field communicating through a transdisciplinary "metaphorical language" or "creole,"
such as theorized by Peter Galison in his notion of scientific "trading zones."
On trading zones as discursive sites where various and otherwise incommensurable knowledge cultures learn to speak to one another, see Galison, *Image
and Logic*. Also see Meyer and Kuusi, "Nanotechnology: Generalizations."

48 On professional confrontations within nanotechnology, see Rotman, "Will the
Real Nanotech Please Stand Up?" On the incommensurability of certain nano
research programs—for example, disciplinary conflicts between chemists and
engineers, as instantiated in the debate between Smalley and Drexler regarding
the feasibility of nanomachines—see Bueno, "The Drexler-Smalley Debate";
Bensaude-Vincent, "Two Cultures of Nanotechnology?"; and Kaiser, "Drawing
the Boundaries of Nanoscience." For related discussion of incommensurability
among different epistemic cultures on the topic of nanoscience, see Wald,
"Working Boundaries on the *nano* Exhibition."

49 See Glimell, "Grand Visions and Lilliput Politics." While such debates succeed
in establishing common terminology and concepts essential to nanoscience—
such as "self-replication"—they also construct variant allegiances to these
concepts as a way of further hardening what counts as "real nanotech." For a
broader account of struggles between scientific programs as the mechanism
for stabilizing disciplinary unity—such as when one research program appropriates or absorbs the theoretical, experimental, and technological resources or
concerns of competitor programs—see Lenoir, *Instituting Science*.

50 Drexler, *Engines of Creation*, 241.

51 On textbooks as the closure or concretization of formerly mobile research
knowledge, see T. Kuhn, *The Structure of Scientific Revolutions*, 20–21, 136–39;
and Traweek, *Beamtimes and Lifetimes*.

52 Schummer, "Interdisciplinary Issues in Nanoscale Research."

53 See, for example, Roco and Bainbridge, *Converging Technologies for Improving
Human Performance*; and Roco and Bainbridge, *Managing Nano-Bio-Info-Cogno
Innovations*. On nanotech's convergence even with the work of contemporary
poetry at the "limits of fabrication," see Nathan Brown, "Needle on the Real."
On the strong utopianism informing claims about nanoscale convergence,
see Saage, "Konvergenztechnologische Zukunftsvisionen und der klassische
Utopiediskurs."

54 Wolfe, "Top Nano Products of 2005," 1.

55 Smalley, foreword to *Nanocosm*, vii.

56 This strategy depends somewhat on the lack of definitional precision in the

term "nanotechnology." Many researchers adopt the term for work in nano-electronics, nanocomposition, or nanolithography that does not necessarily match Drexler's original definition of nanotechnology as molecular manufacturing. On the problems and politics of definition in nanotechnology research, see Decker, Fiedeler, and Fleischer, "Ich sehe was, was Du nicht siehst . . . zur Definition von Nanotechnologie"; and Decker, "Eine Definition von Nanotechnologie."

57 Eigler, quoted in Rotman, "Will the Real Nanotech Please Stand Up?" 53.

58 Reed, quoted in Rotman, "Will the Real Nanotech Please Stand Up?" 48 (emphasis added).

59 An audio recording of Feynman's talk was transcribed, edited slightly, and published in *Engineering and Science* 23 (February 1960): 22–36. A shorter version was published as "The Wonders That Await a Micro-Microscope," *Saturday Review* 43 (1960): 45–47. Feynman released the full transcript in 1961 for Gilbert's *Miniaturization* volume (all cited page references are to this version). It has since been reprinted in numerous scientific journals, as well as in collections of Feynman's writings. It is also available at the Caltech website and other Internet locations. These several incarnations of Feynman's talk, each with independent legacies of citation, suggest the eminence of the speech within the technoscape.

60 Feynman, "There's Plenty of Room at the Bottom," 295.

61 A few examples among thousands: Drexler, "Molecular Engineering," 5275; Drexler, *Engines of Creation*, 40–41; Davis, "Synthetic Molecular Motors," 120; Gimzewski and Joachim, "Nanoscale Science," 1683; Merkle, "Nanotechnology and Medicine," 277–86; Merkle, "Letter to the Editor," 15–16; Sargent, *Dance of the Molecules*, xv–xvi. On the history and legacy of this talk in nanoscience, see Regis, *Nano*, 10–12, 63–94. See also Toumey, "Apostolic Succession." Toumey shows that citations of Feynman's talk in the nanoliterature have been retroactive, a function of rediscovery and genealogical re-creation rather than direct intellectual succession. See also Junk and Riess, "From an Idea to a Vision."

62 Hall, "Overview of Nanotechnology"; Sound Photosynthesis, "Richard Feynman."

63 Booker and Boysen, *Nanotechnology for Dummies*, 13 (ellipses in original).

64 Merkle, "Response to *Scientific American*."

65 Feynman's reputation for uncanny scientific insight, his prodigious "genius," was legendary in his own lifetime and has continued to grow; see Gleick, *Genius*.

66 Drexler's 1992 testimony is found in the U.S. Senate report *New Technologies for a Sustainable World* (1993). The 1999 testimonies are reported by the U.S. House of Representatives in *Nanotechnology: The State of Nanoscience and Its Prospects for the Next Decade* (1999). Federal documentation for the National Nanotechnology Initiative is compiled by the National Science and Technology Council in *National Nanotechnology Initiative* (2000). For an excellent history

of the NNI and the role of utopian visions in implementing funding for nano-science, see McCray, "Will Small Be Beautiful?"

67 Clinton, address to the California Institute of Technology.

68 Drexler, "From Nanodreams to Realities." Drexler's argument is similar to that of Arthur C. Clarke in his *Profiles of the Future*. Frequently cited in Drexler's publications, Clarke's text aggressively foregrounds the science fiction founda-tions of scientific extrapolation.

69 Drexler, *Engines of Creation*, 234–35.

70 Baudrillard, "The Precession of Simulacra," 1–42.

71 Heinrich et al., "Molecule Cascades," 1381.

72 As of 2008, the expanded Science Citation Index records that Drexler's *Nano-systems* has been cited more than 500 times in the technical scientific literature, while *Engines of Creation* has been cited more than 200 times, and the 1981 PNAS article "Molecular Engineering" has been cited more than 175 times. This cita-tion history demonstrates a substantial impact, to say the least. We should also note the technology companies founded in the 1990s to pursue some aspect of Drexler's vision, especially Zyvex. On Zyvex's efforts, see Ashley, "Nanobot Construction Crews."

73 Smalley, quoted in Voss, "Moses of the Nanoworld," 62.

74 Smalley, "Smalley Responds," 39. For many years, Smalley fervently supported Drexler's vision and participated in several meetings of the Foresight Insti-tute. In 1993, while spearheading Rice University's nanotechnology initiative, Smalley said:

> I am a fan of Eric. . . . I am a fan of his, and in fact in my endeavors to explain to people what I thought the future was, particularly the board of governors here at Rice, I have given them copies of some of Eric's books. . . . What I like about Eric is his very vivid imagining of how exciting this future could be . . . I'm very curious about this. I'd love to build one of these things, so when I read what he's talking about, I'm ready to do it.
>
> Somebody's gonna do this someplace. I'd dearly love it to be us. (Quoted in Regis, *Nano*, 273–77)

However, by the time the NNI launched, Smalley's allegiance with Drexler had deteriorated. While Smalley's congressional testimony in 1999 remained en-thusiastic about a generally Drexlerian future where "we learn to build things at the ultimate level of control, one atom at a time," making several references to Drexler's "delightful" books, he also asserted that the specific idea of a univer-sal assembler "will always remain a fantasy. It is just a dream" (U.S. House of Representatives, *Nanotechnology: The State of Nanoscience*, 55, 76–77). By 2001, Smalley began to publish his opinion that Drexlerian assemblers would be im-possible to produce in reality and that nanotech would develop along different lines; see Smalley, "Of Chemistry, Love and Nanobots."

75 López, "Bridging the Gaps." López shows that science-fictional narrative ele-

ments structure mainstream nanodiscourse—including publications of the National Science Foundation on NBIC convergence—in the same way as Drexler's *Engines of Creation*. Also see Grunwald, "Nanotechnologie als Chiffre der Zukunft."

76 See Regis, *Nano*, 3–18.

77 Merkle, "Letter to the Editor," 15.

78 Culler, *The Pursuit of Signs*, 38.

79 Stephenson, *The Diamond Age*, 41–42.

80 Yakobson and Smalley, "Fullerene Nanotubes," 332. Clarke's rendition of the space elevator itself borrows from a long history of speculative science on the topic; see Clarke, "The Space Elevator." Although Clarke depicted the elevator in *The Fountains of Paradise* as made from diamond, he now follows Smalley in believing that, were it ever actually to be built, fullerenes would be the material of choice. Given the many advances in fullerene nanotechnology, Clarke has also recently revised his original estimates for the elevator's arrival: "As its most enthusiastic promoter, I am often asked when I think the first space elevator might be built. My answer has always been: about 50 years after everyone has stopped laughing. Maybe I should now revise it to 25 years" ("1st Floor," 25).

81 On the role of nanotech in the conflict between humanism and posthumanism in *Star Trek*, see Relke, *Drones, Clones, and Alpha Babes*.

82 Drum and Gordon, "Star Trek Replicators and Diatom Nanotechnology," 327. The authors here refer to the research of Sandhage et al., "Novel, Bioclastic Route to Self-Assembled, 3D, Chemically Tailored Meso/Nanostructures." They also cite Okuda et al., *The Star Trek Encyclopedia*.

83 Hall, "Utility Fog," 161–84.

84 Hall, *Nanofuture*, 195.

85 Ibid., 189. On the broader relationship between nanoscience and superhero fiction, see C. Milburn, "Nanowarriors."

86 For example, see the following nanowritings: Lee et al., "Nanosignal Processing"; Kosko, *Fuzzy Thinking*, 250–95; Kosko, *The Fuzzy Future*, 133–34, 240–56; Slonczewski, "Tuberculosis Bacteria Join UN"; McCarthy, *Hacking Matter*; McCarthy, "Programmable Matter"; Benford, "BIO/NANO/TECH"; Benford, "Calculating the Future"; Benford, "A Diamond Age"; Benford and Malartre, *Beyond Human*; Brin, "Singularities and Nightmares"; Cramer, "Report on Nanocon 1"; Cramer, "The Carbon Nanotube"; and Cramer, "Nanotechnology: The Coming Storm." Kosko, Benford, Brin, and Cramer have all had involvement with the Foresight Institute, and Brin has served on the Center for Responsible Nanotechnology's Global Implications and Policy Task Force since 2005.

87 For example, see the following nanofictions: Kosko, *Nanotime*; Slonczewski, *Brain Plague*; McCarthy, *Murder in the Solid State, Bloom, The Collapsium*, and *The Wellstone*; Benford, *Great Sky River, Tides of Light, Furious Gulf*, and *Sailing Bright Eternity*; Brin, *Earth*; and Cramer, *Einstein's Bridge*.

88 Slonczewski, "Stranger than Fiction," 501.

89 Bainbridge, *Dimensions of Science Fiction*, 151. Cf. Bainbridge, "The Impact of Science Fiction" and *The Spaceflight Revolution*.

90 Bainbridge, "Sociocultural Meanings of Nanotechnology."

91 Bainbridge, "Religions for a Galactic Civilization."

92 Bainbridge, *Nanoconvergence*; Roco and Bainbridge, *Converging Technologies for Improving Human Performance*; Roco and Bainbridge, *Managing Nano-Bio-Info-Cogno Innovations*; Roco and Bainbridge, *Nanotechnology: Societal Implications*.

93 Bainbridge, *Nanoconvergence*, 34–35, 46–47. Bainbridge here reworks points he makes in "Sociocultural Meanings of Nanotechnology," 297–98. For a further account of the use value of science fiction in shaping attitudes towards nanotechnology, see Berne and Schummer, "Teaching Societal and Ethical Implications of Nanotechnology."

94 Theis, "Letter to the Editor," 15.

95 See Regis, *Nano*, 152–54.

96 Heinlein, "Waldo," 29, 133. Heinlein originally published "Waldo" in *Astounding Science Fiction* (August 1942) under the pseudonym Anson MacDonald.

97 Feynman, "There's Plenty of Room at the Bottom," 292.

98 Heinlein, "Waldo," 133 (emphasis added). Heinlein's character of Waldo—brain surgeon, inventor, professional dancer, and discoverer of the scientific basis of magic—was probably modeled after a friend of Heinlein's whose own life seems to have unfolded according to "science (fiction)" ways of seeing: John Whiteside Parsons. Parsons was one of the founders of the experimental rocket research group at Caltech, helping to transition spaceflight from science fiction to real science. He was among the earliest members of the Jet Propulsion Laboratory (and his lingering reputation there was surely known to Albert Hibbs, who joined JPL in 1950). Parsons was also a science fiction enthusiast actively engaged with the California sf community. But most exceptionally, he was a devoted practitioner of occult magic who wanted to establish a technical foundation for working with supernatural forces. He died in a mysterious chemical explosion in 1952, which drew significant media attention to his colorful history. See Pendle, *Strange Angel*, esp. 228–31. See also Carter, *Sex and Rockets*. Heinlein's "Magic, Inc." (aka "The Devil Makes the Law," 1940), published in book form together with "Waldo" originally in 1950, also bears strong traces of Parsons, as does the later *Stranger in a Strange Land* (1961). It is tempting to see nanotech's aura of the magical, of the impossible made real, as carried through the Parsons-Heinlein-Hibbs-Feynman genealogy.

99 Derrida, "Structure, Sign, and Play in the Discourse of the Human Sciences," 292. With a similar design to denaturalize the foundations of humanism, Michel Foucault has written, "As the archaeology of our thought easily shows, man is an invention of recent date. And one perhaps nearing its end" (*The Order of Things*, 387). It is worth noting that Derrida, however, critiques claims for the end of man by Foucault and others as a reinscription of humanist eschatology. Although ultimately affirming the necessity of Foucault's move out-

side the boundaries of humanism, Derrida suggests that this strategy must be accompanied by deconstruction from within to become other than another sedimented humanism; see Derrida, "The Ends of Man."

100 For some examples of posthuman phenomenological spaces created at the intersections of bodies and technologies, see Virilio, *The Art of the Motor*; Massumi, *Parables for the Virtual*; Stone, *The War of Desire and Technology at the Close of the Mechanical Age*; Hayles, *How We Became Posthuman*; Hayles, "Flesh and Metal"; Hansen, *Embodying Technesis*; and Waldby, *The Visible Human Project*. On hybrid-reality spaces that integrate the human body with the nano-imaginary, see de Souza e Silva, "The Invisible Imaginary." I discuss the phenomenological zones that form at the flesh-machine interfaces of nanoscience more thoroughly in chapter 2.

101 Hurley, "Reading like an Alien," 220.

102 On thematic characteristics of nanonarratives, see Landon, "Less Is More"; and Attebery, "Dust, Lust, and Other Messages from the Quantum Wonderland."

103 Extropians maintain that human life is rapidly evolving through the mediation of emerging technosciences and the active pursuit of science fiction scenarios, such as "uploading" and cryonic suspension. Max More, founder of the Extropy Institute, writes that extropians "advocate using science to accelerate our move from human to a transhuman or posthuman condition," and that nanotechnology is a significant vehicle in bringing about the posthuman future; see More, "The Extropian Principles 3.0." Nanotechnologists like Merkle, Drexler, and Kosko have been active with the Extropy Institute and other transhumanist organizations. Hall has served as the nanotechnology editor of *Extropy: The Journal of Transhumanist Solutions*. Bainbridge too has been a booster for transhumanism; see his *Across the Secular Abyss*, "Cyberimmortality," and "Progress toward Cyberimmortality." For several years he has also been developing "personality capture" systems, envisaged as a prelude to uploading; see Bainbridge, "Massive Questionnaires for Personality Capture"; and "A Question of Immortality." On the function of transhumanism in the social development of nanotechnology, see Coenen, "Der posthumanistische Technofuturismus in den Debatten über Nanotechnologie und Converging Technologies."

104 On forms of posthumanism that delete the body in favor of a fantasy of disembodied information or immaterial transcendence, see Hayles, *How We Became Posthuman*, esp. 1–24; Springer, *Electronic Eros*, 16–49; Dery, *Escape Velocity*, 229–319; and Doyle, *Wetwares*, 43–145. These theorists instead point to ways of reimagining posthumanism through a fundamental commitment to materiality, recognizing that it is possible to think outside obsolete and disenabling forms of humanism without rejecting the site of our being in the world, namely, the body. See also Halberstam and Livingston, *Posthuman Bodies*; Graham, *Representations of the Post/Human*; and Vint, *Bodies of Tomorrow*.

105 Montemagno, "Nanomachines," 1.

106 Drexler, *Engines of Creation*, 38.

107 Feynman, "There's Plenty of Room at the Bottom," 295.

108 Merkle, who left Xerox in 1999 to become principal fellow of Zyvex, and later, in 2003, professor at Georgia Tech College of Computing, jokes about the impact of copy culture on nanotech in "Design-Ahead for Nanotechnology," 35.

109 Drexler, *Engines of Creation*, 103.

110 The fantasy of "telegraphing a human" appears frequently within posthuman science (fiction): for example, Wiener, *Human Use of Human Beings*, 95–104; and Moravec, *Mind Children*, 116–22. Hall discusses nanotech bodily telegraphy in *Nanofuture*, 169–70.

111 The "gray goo" hypothesis imagines the entire organic world dismantled into disorganized material by rampaging nanobots: Drexler, *Engines of Creation*, 172–73; Regis, *Nano*, 121–24. In 2000 this horrifying possibility so disturbed Bill Joy, cofounder and former chief scientist of Sun Microsystems (and certainly no technophobe), that in a now famous article he questioned the wisdom of pursuing nanoresearch, suggesting that the future would perhaps be a better place if we did not follow our nanodreams. See Joy, "Why the Future Doesn't Need Us." Gray goo will be considered more thoroughly in chapter 3.

112 For a sustained discussion of nanotechnology's significance for, and commitment to, cryonics, see Du Charme, *Becoming Immortal: Nanotechnology, You, and the Demise of Death*. In Du Charme's account—typical of nanowriting—"you" are already a posthuman subject of the future. For analysis of these discourses where cryonics and nanotechnology come together in producing a suspended subject, strung out between the present and the future, see Doyle, *Wetwares*, 63–118, 136–41.

113 The Alcor Foundation, founded in 1972, promotes public awareness of cryonic possibilities and assists its members in arranging for cryonic suspension after their deaths. Alcor cites fully functional nanotechnology as the fundamental scientific development still needed to repair and resurrect suspended patients. See the Alcor Life Extension Foundation website, www.alcor.org. The Extropy Institute, in which many nanoscientists are involved, traces its historical origins to the Alcor Foundation, signaling the deeply complicit nature of posthuman discourse and cryo-nanotechnology. For more on the interlinkage of cryonics, nanotechnology, and posthuman immortality, see Regis, *Great Mambo Chicken and the Transhuman Condition*, 1–9, 76–143. See also Merkle, "Cryonics."

114 Drexler, "Molecular Engineering," 5278.

115 Drexler, *Engines of Creation*, 136–38.

116 No cryonics institute admits to holding Walt Disney's body, despite the widespread cultural awareness, or myth, of Walt Disney on ice (and of course I mean both the frozen man and the annual "Disney on Ice" skating extravaganzas produced across North America by the Walt Disney Company). So it may be that Disney is not on ice, or it may be that the cryonics institute which may (or may not) have Disney in storage (is it Alcor? or not?) is simply following a standard secrecy policy for protecting client privacy. In 1966, Disney's family members

refused to comment on the details of his death, but they have since publicly insisted that he was cremated—indeed, an urn with Walt Disney's name on it seems to be interred at Forest Lawn Cemetery in California—though conspiracy theorists see the Disney burial plot as nothing more than a ruse. The truth of Walt Disney on ice can neither be confirmed nor be denied with absolute certainty. But such a paradox only participates in the suspension of the cryonic subject between being and nonbeing, an endless undecidability: Disney both is and is not frozen, he both is and is not dead, he both is and is not anywhere at all. See Doyle, *Wetwares*, 68–70.

117 Baudrillard, *Écran total*, 169–73; Baudrillard, "The Precession of Simulacra," 12–14. For analysis of Disneyism and its production of posthuman "terminal identities," see Bukatman, *Terminal Identity*, 227–29, and "There's Always Tomorrowland."

118 McKendree, "Nanotech Hobbies," 143.

119 These NSF-funded nano exhibitions were developed by Cornell University together with the Sciencenter, Painted Universe, and other collaborators; see "It's a Nano World," http://www.sciencenter.org/nano; and "Too Small to See: Zoom into Nanotechnology," http://www.toosmalltosee.org.

120 Merkle, "It's a Small, Small, Small, Small World," 26. Cf. Regis, *Great Mambo Chicken*, 2, 126–30.

121 Merkle, "It's a Small, Small, Small, Small World," 32. Merkle actually cautions that nanotech will *not* develop on its own, but then strangely implies that even if we do not strive for the nanofuture, if "we ignore it, or simply hope that someone will stumble over it," the evolution of nanotech will still proceed—it is still inevitable—but just "will take much longer."

122 Stephenson, *The Diamond Age*, 31.

2. SMALL WORLDS

1 Freud, "The Uncanny," 241.

2 McLuhan, *Understanding Media*, 5.

3 Jameson, *Postmodernism*, 412–13.

4 Hissink, "Nanotechnology Makes a Small World Even Smaller."

5 Short, "How Small Is Small?"

6 Michalske, quoted in Pressman, "Nano Narrative," 192.

7 Brezinski, "It's a Small World after All," 12.

8 Ibid.

9 Harrow, "A Hundred Billion Billion Bytes" (ellipses in original).

10 Smarr, "Microcosmos: Nano Space," 134.

11 G. Milburn, *Schrödinger's Machines*, 180.

12 Cover headline for Bruno, "The Next Big Thing Is Incredibly Tiny."

13 National Science and Technology Council, *Nanotechnology: Shaping the World Atom by Atom*, 1.

14 G. Milburn, *Schrödinger's Machines*, i.

15 Galison, *Image and Logic*, 52.

16 The tropic protocol, then, would be one of the many forms of knowledge that become integrated in scientific instruments as what Davis Baird calls "thing knowledge." Baird writes that a scientific instrument "is encapsulated thing knowledge synthesizing working knowledge, model knowledge, theoretical knowledge, and functional equivalents to skill knowledge" (*Thing Knowledge*, 88). The instrument brings together these forms of knowledge with material reality: "The material medium of the instrument encapsulates and integrates all these different kinds of knowledge. All are necessary for the instrument to render information about a specimen" (70). The tropic protocol in these terms would be a kind of rhetorical knowledge, or cultural narrative knowledge, that adds another aspect to synthetic thing knowledge, and it affects the understanding of rendered information about a specimen as much as any other form of knowledge encapsulated by the instrument. For a related account of the entangled economies of rhetorical strategies, textual and visual tropes, and the functionality of scientific instruments, see Biagioli, *Galileo's Instruments of Credit*.

17 For example, as Jochen Hennig shows, the technical design of STM images changed over the course of a decade in adapting to the nanorhetoric of atoms as "building blocks" and visions of "shaping the world atom by atom"; see Hennig, "Changes in the Design of Scanning Tunneling Microscope Images from 1980 to 1990."

18 Eigler writes: "Atom manipulation came about almost by accident. . . . In September of 1989, Erhard Schweizer and I were continuing these studies [of xenon atoms on a platinum surface] when we noted some unusual streaks across our STM images which were clearly due to tip induced motion of the xenon adatoms. . . . We learned that the presence of the streaks depended on how we operated the microscope. If we operated the microscope with the tip close enough to the sample then we saw the streaks, otherwise we did not. This immediately suggested that we could use the tip to control the position of the xenon" (Eigler, "From the Bottom Up," 430–31).

19 Ibid., 427.

20 Eigler and Schweizer, "Positioning Single Atoms," 525.

21 Ibid., 524.

22 Ibid., 526.

23 Crommie, Lutz, and Eigler, "Confinement of Electrons to Quantum Corrals on a Metal Surface," 218.

24 IBM Almaden Visualization Lab, "The Corral Reef." On interpretations of quantum mechanics in nanoscience, see Vermaas, "Nanoscale Technology."

25 Mapping practices of various sorts have historically contributed to the transformation of physical properties into property as such. On geographic mapping and its literary accomplices as working to secure political identities and spatial

arrangements, see Conley, *The Self-Made Map*. On literary tools of mapping—for example, the blazon, which inventories new land simultaneously as it catalogs the female body—and the rhetoric of property that establishes order by drawing proper boundaries, see Parker, *Literary Fat Ladies*. On electronic worldmapping as facilitating the scientific "disciplining of time," see Galison, *Einstein's Clocks, Poincaré's Maps*. On "data mapping" specifically, as a form of visualization that enframes, and tames, large data sets into maps more easily digestible by human scales, see Manovich, "The Anti-sublime Ideal in Data Art."

26　Valerie Hanson has shown that nanodiscourse, in producing a sense of place in the nanoscale and in constructing the "nanoworld" as a shared locality of STM research, relies on an inherited tropology of "new worlds" and "invisible worlds" (and all the exploratory and frontier associations of these tropes) going back to the Scientific Revolution; see V. Hanson, "Haptic Visions," chap. 4. Alfred Nordmann has argued that nanotechnology is a "place-oriented enterprise" driven to occupy and stake claims in the nanoworld. Unlike theory-driven scientific research or device-driven technology research, nanotechnology should be considered "exploratory research": "Research that aims for conceptual as well as physical mastery of a certain territory, domain, or size regime, is interested neither in theory nor merely in novel devices and substances. It is exploratory research, literally speaking, where settlement follows upon exploration and new practices, perhaps a new culture is founded" ("Molecular Disjunctions," 59). Also see Anderson, Kearnes, and Doubleday, "Geographies of Nanotechnoscience."

27　Timp, "Nanotechnology," 2.

28　G. Milburn, *Schrödinger's Machines*, 1.

29　National Science and Technology Council, *Nanotechnology: Shaping the World*, 4.

30　Heidegger, "Age of the World Picture," 134.

31　Ibid., 135.

32　The phrase "the limits of fabrication"—sometimes "the ultimate limits of fabrication"—appears ubiquitously in nanowritings to indicate that nanotechnology takes our mastery of materiality to the climax of what is theoretically possible in accord with the laws of physics. See, for example, Welland and Gimzewski, *The Ultimate Limits of Fabrication and Measurement*; and Drexler, "Molecular Manufacturing: Perspectives on the Ultimate Limits of Fabrication."

33　See Nathan Brown, "Needle on the Real." Brown suggests that the fabrications of protosematic poetry and STM atomic manipulations, both engaging the materialities of signification at, or passing over, their limits, point beyond the structural systems in which they may signify (that is, beyond the dimensions of the human and the symbolic), toward possibilities of ethical encounter with the real.

34　Binnig and Rohrer, "In Touch with Atoms," S324.

35　Binnig and Rohrer, "Scanning Tunneling Microscopy," 624.

36 For example, Hameroff, *Ultimate Computing*, 190–215; and Schneiker et al., "Scanning Tunnelling Engineering." Schneiker seems to have been the first scientist to make an explicit connection between the STM and a "Feynman machine." Later uses of the analogy can be traced back to Schneiker's series of unpublished manuscripts from the 1980s via various direct and indirect genealogies; see Toumey, "Reading Feynman into Nanotech"; and Toumey, "The Man Who Understood the Feynman Machine."

37 Eigler, "There's Plenty of Room in the Middle."

38 The machinations of the "already inevitable" in the historical reconstruction of nano exemplify what Erich Auerbach famously described as the "figural interpretation of history" endemic to Christian metaphysics, which "establishes a connection between two events or persons in such a way that the first signifies not only itself but also the second, while the second involves or fulfills the first. The two poles of a figure are separated in time, but both, being real events or persons, are within temporality. . . . The here and now is no longer a mere link in an earthly chain of events, it is simultaneously something which has always been, and which will be fulfilled in the future" (Auerbach, *Mimesis*, 73–74). But where the collapse of temporality within the Christian figural interpretation of history relies, as Auerbach shows, on a different history outside history (that is, on the presence of divinity outside time, beyond the end of history), the nano-technological figural interpretation of history and its collapse of temporality operate by evacuating the beyond, such that there is no longer any history outside history, for the outside will be found already inside. We will soon see how this techno-deconstructive effect of nanovision works.

39 Eigler, quoted in Appenzeller, "The Man Who Dared to Think Small," 1300. On the "anxiety of influence" as one creative individual's productive and conflicted engagement with an earlier (and usually dead) visionary, see Bloom, *The Anxiety of Influence*. On the way in which intellectual genealogies are constituted as an experience of inheritance, legacy, possession, or haunting only belatedly, as a "post effect" where the past inherits from the present as much as the converse, see Derrida, *The Post Card* and *Specters of Marx*.

40 Drexler, *Engines of Creation*, 240.

41 National Science and Technology Council, *Nanotechnology: Shaping the World Atom by Atom*, 6, 7.

42 Mody, "How Probe Microscopists Became Nanotechnologists." The STM plays this central role in the "standard story" of nanotechnology's development (despite the many difficulties of producing STM images relative to other forms of microscopy capable of nanoscale resolution) because of its ability to manipulate as well as visualize atoms, an ability which, while quite limited at present, helps to reinforce the visionary predictions of nanofuturology. In the standard story, the STM is made (perhaps erroneously) to represent something like a "proto-assembler"; see Baird and Shew, "Probing the History of Scanning Tunneling Microscopy."

43 Feynman, "There's Plenty of Room at the Bottom," 286, 296.

44 Ibid., 295.

45 Mamin et al., "Gold Deposition from a Scanning Tunneling Microscope Tip."

46 National Science and Technology Council, *Nanotechnology: Shaping the World Atom by Atom*, 2.

47 Borges, "On Exactitude in Science" (1946).

48 For optical microscopy, the Rayleigh limit of resolution is given by a constant (0.61) multiplied by the wavelength of light, divided by the objective numerical aperture (that is, the minimum resolvable distance between two point sources is approximately $0.61 \ \lambda/n_a$). Using the shortest wavelength of visible light (blue: $\lambda = 450$ nm), and the best available high-res lenses, the spatial resolution turns out to be just under 200 nm (thus precluding observation of most single-molecule phenomena). Some forms of optical microscopy, for example, scanning near-field optical microscopy (SNOM), are able to surpass the Rayleigh limit and achieve resolution as good as 50 nm. Significantly, however, this is by virtue of the kind of point-by-point scanning and near-field closeness similar to that of scanning probe microscopes. Like electron microscopes under the best of conditions, the STM and other probe microscopes can achieve atomic resolution. But more important than the fact of atomic resolution, these forms of microscopy that bring otherwise impossible objects to light within near-field closeness, or what I will later discuss as (Deleuzian) "haptic space," materialize the tropic protocol of "touch" and "no distance" that determines the small-world effect of nanovision. But I am getting ahead of myself . . .

49 G. Milburn, *Schrödinger's Machines*, 47.

50 Virilio, *War and Cinema*, 88.

51 Irigaray, interview in *Les femmes, la pornographie, l'érotisme*, ed. Marie-Françoise Hans and Gilles Lapouge (Paris: Seuil, 1978), 50, quoted in Jay, *Downcast Eyes*, 493.

52 Haraway, "Situated Knowledges," 188.

53 National Science and Technology Council, *Nanotechnology: Shaping the World Atom by Atom*, 1, 8, i.

54 Smalley, "Nanotechnology: Prepared Written Statement," 55.

55 See Nordmann, "Nanotechnology's Worldview." Nordmann notes how such imagistic overlaps of the nanocosm with the macrocosm, the molecular with the cosmic, help to affirm nanotech's characteristic collapse of nature and technology. For more on the way in which nanotechnology's worldview (a.k.a. nanovision) problematizes traditional divisions between nature and technology, see Schiemann, "Dissolution of the Nature-Technology Dichotomy?"; and Köchy, "Maßgeschneiderte nanoskalige Systeme."

56 See Turner, *The Frontier in American History*; and Hardt and Negri, *Empire*.

57 Lacan, *Seminar of Jacques Lacan, Book I*, 66.

58 Žižek, *The Sublime Object of Ideology*.

59 Eigler, "From the Bottom Up," 426–27.

60 Committee for the Review of the National Nanotechnology Initiative, *Small Wonders, Endless Frontiers*.

61 Bush, *Science, the Endless Frontier*. On the history of American technoscientific development in the modern era as driven by the fugitive sublime, see Nye, *American Technological Sublime*: "Despite its power, the technological sublime always implies its own rapid obsolescence, making room for the wonders of the next generation. . . . Each form of the technological sublime [in American history] became a 'natural' part of the world and ceased to amaze, though the capacity and the desire for amazement persisted" (283–84).

62 Baudrillard, *The Illusion of the End*, 112.

63 Rheinberger, *Toward a History of Epistemic Things*, 23.

64 Nordmann, "*Noumenal* Technology."

65 Miniaturization in general has traditionally facilitated nostalgic objectification and desire for domestic containment; see Stewart, *On Longing*. Nanotechnology is clearly no exception to this history of miniaturization.

66 In "What's the Buzz?" Susan Lewak shows how Carroll's *Alice in Wonderland* has frequently served as a metaphor in writings about nano. As in Carroll's text, the wonder initially experienced in the scientific encounter with the "bizarre world of the quantum" diminishes as the rules and laws of its operation are learned. Like Alice, nanoscience eventually finds little wonder in wonderland.

67 See Hessenbruch, "Beyond Truth"; and Landon, "Less Is More."

68 As the science fiction author and critic Damon Knight famously suggested, the location of wonder is in the "impossible," the space made by a "widening of the mind's horizons" and increasing the distance between the real and the imaginary: "Science fiction exists to provide . . . 'the sense of wonder': some widening of the mind's horizons, no matter in what direction . . . any new sensory experience, impossible for the reader in his own person, is grist for the mill, and what the activity of science fiction writing is all about" (*In Search of Wonder*, 24). For more on the "sense of wonder" and its role in genre sf, see Landon, *Science Fiction after 1900*, 18–20.

69 Ball, "It's a Small World," 30.

70 Binnig and Rohrer, "Scanning Tunneling Microscopy," 623. See also Hessenbruch, "Nanotechnology and the Negotiation of Novelty." Hessenbruch shows that the novelty of probe microscopy could only be established relative to a simultaneous discourse of banality, a constant negotiation of the legitimately new with its similarity to the old. Andreas Lösch, in "Nanomedicine and Space," has also argued for this dynamic between discursive orders of radical innovation and familiar continuity in mediating the revolutionary aspects of nanoscience. Also see Lösch, "Means of Communicating Innovations."

71 Campbell, *Wonder and Science*, 182.

72 Marton, "Alice in Electronland." Nicolas Rasmussen provides an excellent account of the history of electron microscopy, Marton's role in making electron microscopy viable for biological research, and the didactic value of Marton's

imaginative "Alice" article in this effort; see N. Rasmussen, *Picture Control*, 41–43. Like depictions of "wonderland" in nanodiscourse, Marton's rendition of "Electronland" finds the small world a place of altered perception—"The world was singularly changed" ("Alice in Electronland," 247)—but the weirdness and wonderment rapidly diminish as this world becomes increasingly explained and pictured.

73 N. Rasmussen, *Picture Control*, 27.

74 In "The Work of Art in the Age of Mechanical Reproduction," Walter Benjamin described the age of mechanical reproduction as the disappearance of the sublime—the "aura" attendant to the concept of the "original" now lost by the replicability of representation.

75 On the policing of wonder by Enlightenment and post-Enlightenment experimental science, see Daston and Park, *Wonders and the Order of Nature*. On scientific inquiry as a penetration into the unseen, pursuing rationality by invading the holdouts of the sublime, see Stafford, *Body Criticism*. On the pornographic as an aspect of scientific visual penetration into the unseen, see Williams, *Hard Core*.

76 Benjamin, "The Work of Art in the Age of Mechanical Reproduction," 222.

77 Manovich, "The Anti-sublime Ideal in Data Art." Manovich writes that the "promise of rendering the phenomena that are beyond the scale of human senses into something that is within our reach, something visible and tangible . . . makes data mapping into the exact opposite of the Romantic art concerned with the sublime. . . . If Romantic artists thought of certain phenomena and effects as un-representable, as something which goes beyond the limits of human senses and reason, data visualization projects aim at precisely the opposite: to map such phenomena into a representation whose scale is comparable to the scales of human perception and cognition."

78 The STM therefore cannot be said to let us "see" at the nanoscale, since it does not really reproduce a mimetic representation of the dimensional relationships present in the sample so much as re-create them by algorithmic manipulation; see Robinson, "Images in Nanoscience." As Hacking has written of our ability to see at microscopic scales: "When an image is a map of interactions between the specimen and the image of radiation, and the map is a good one, then we are seeing with a microscope. What is a good map? After discarding or disregarding aberrations or artifacts, the map should represent some structure in the specimen in essentially the same two- or three-dimensional set of relationships as are actually present in the specimen" (*Representing and Intervening*, 208). The STM does not, then, in this sense make "good maps" or let us "see" as a technological extension of normal, optical vision; see Pitt, "Epistemology of the Very Small" and "When Is an Image Not an Image?" On the problems of seeing and shaping algorithmic scientific images through enculturated assumptions of mimesis, see Dumit, *Picturing Personhood*.

79 Gerber and Lang, "How the Doors to the Nanoworld Were Opened," 4.

80 Binnig and Rohrer, "In Touch with Atoms," S325; cf. Binnig et al., "7 X 7 Reconstruction on Si (III) Resolved in Real Space."

81 Binnig and Rohrer, "Scanning Tunneling Microscopy," 620.

82 Gimzewski and Vesna, "The Nanomeme Syndrome," 14.

83 Brauman, "Room at the Bottom," 1277.

84 G. Milburn, *Schrödinger's Machines*, 45–46.

85 Simon, "Manipulating Molecules," 36.

86 Ibid. The book collection *Waldo and Magic, Inc.* was actually first published in 1950.

87 Robinett, quoted in Sincell, "NanoManipulator Lets Chemists Go Mano a Mano with Molecules," 1530.

88 Binnig and Rohrer, "In Touch with Atoms," S324.

89 Deleuze, *Francis Bacon*, 125, 99. Deleuze adapts Riegl's discussion of haptic vision; see Riegl, *Late Roman Art Industry*.

90 Deleuze and Guattari, *A Thousand Plateaus*, 494.

91 V. Hanson, "Haptic Visions."

92 Eigler, "From the Bottom Up," 427.

93 Binnig and Rohrer, "In Touch with Atoms," S324.

94 Ibid., S325. On sensory transformations of STM data, see Soentgen, "Atome Sehen, Atome Hören."

95 Stroscio and Eigler, "Atomic and Molecular Manipulation with the Scanning Tunneling Microscope," 1319.

96 Bauer and Rogers, "A Billionth of a Meter Is a Big Deal."

97 Gimzewski and Vesna, "The Nanomeme Syndrome," 7.

98 Nanovision would therefore be a form of what John Johnston calls "machinic vision"; see Johnston, "Machinic Vision." Unlike the autonomous "vision machine" described by Virilio in *The Vision Machine*, 59–77, in which observation has been severed from the human, nanovision refuses a hierarchy or absolute separation between human and machinic processes of perception. On the reinterpretive negotiations required by such machinic vision in nanoscience, see Missomelius, "Visualisierungstechniken."

99 Gimzewski and Vesna, "The Nanomeme Syndrome," 11.

100 On ways in which human operators enter into embodiment relations with technology, see Ihde, *Technology and the Lifeworld* and *Bodies in Technology*. On embodiment relations historically developed between electron microscopists and their instruments, see N. Rasmussen, *Picture Control*, 222–56. For a deeper account of specific modes of interactivity that link operators with scanning probe microscopes in material assemblages, see V. Hanson, "Haptic Visions," chap. 2.

101 Andy Clark, a cognitive scientist and cybertheorist, argues in *Natural-Born Cyborgs* that the human cognitive and sensory apparatus is preadapted to absorb and expand along its technical prosthesis.

102 G. Milburn, *Schrödinger's Machines*, 45.

103 Kittler, *Gramophone, Film, Typewriter*, 16.

104 Binnig et al., "Tunneling through a Controllable Vacuum Gap."

105 Sargent, *The Dance of Molecules*, 20.

106 Bolter and Grusin, *Remediation*, 23–24. For more on immersive immediacy and telepresence in new media, see Manovich, *The Language of New Media*.

107 "The nanoManipulator: A Virtual-Reality Interface to Scanned-Probe Microscopes," 1. This document is based on the original technical report by Taylor et al., "The Nanomanipulator: A Virtual Reality Interface for a Scanning Tunneling Microscope."

108 Hansen, *New Philosophy for New Media*, 21–22.

109 Ibid., 206. See also Massumi, *Parables for the Virtual*.

110 Moreover, from certain quantum mechanical perspectives, the overlapping orbitals of the tip and the sample enabling the tunneling current are better accounted for as one entity, a single "supermolecule" wherein touch and being have become complementary concepts. See Farazdel and Dupuis, "All-Electron *ab Initio* Self-Consistent-Field Study of Electron Transfer in Scanning Tunneling Microscopy." This theoretical model of STM-sample interaction developed by Farazdel and Dupuis at IBM works as follows:

> Tip and sample are treated as a whole and are allowed to adjust to the presence of one another. . . . To account for this relaxation, we treat the entire STM-STS system (i.e., tip, sample, etc.) as a many-body problem (all electrons and nuclei) and think of the system as a supermolecule. Tip and sample are dealt with on equal footing. . . . In the supermolecule approach the STM-STS tunneling is an electron transfer (ET) *within* the supermolecule. Moreover, at large tip-sample separation, the tunneling is mostly "through space" instead of "through bond," since there are no chemical bonds between the tip and the sample to mediate the transfer of the electrons. In other words, the tunneling here is primarily due to the spatial overlap of the wave functions of the system in the initial state of ET with that of the final state of ET. (3909–10)

> In reconceiving the tip-sample coupling as an orbital overlap that produces a supermolecular whole, the "space" through which electrons tunnel is now contained *inside* the system, and the last remaining vestige of the sublime outside is indeed theorized as "*within* the supermolecule" rather than between; intermediary space is captured by the nanovisual apparatus such that now there is no intermediary space, or else all space is intermediary.

111 Binnig and Rohrer, "In Touch with Atoms," S325.

112 Störmer, quoted in National Science and Technology Council, *Nanotechnology: Shaping the World Atom by Atom*, 1.

113 Haraway, "Situated Knowledges," 189.

114 Cummings, *The Girl in the Golden Atom*, 1. The short stories that make up this novel were originally published in *All-Story Weekly*, March 1919 and January–

February 1920, respectively. On the extent to which Cummings's stories reenact features of earlier science fiction texts, including Wells's *Time Machine* (1895) and Fitz-James O'Brian's "The Diamond Lens" (1858)—the latter of which could also be seen as a precursor to nanotechnology fictions—see Mullen, "Two Early Works by Ray Cummings."

115 Heidegger writes that "the essence of technology is by no means anything technological" but is rather the logic of "enframing" that orders the world-at-hand as "standing reserve" ("The Question Concerning Technology," 4).

116 Feldstein and Gaines, "Lost in the Microcosm," 14; hereafter cited in the text.

117 Matheson, *The Shrinking Man*, 206–7. The documentary film, *Powers of Ten* (dir. Charles and Ray Eames, 1977), also uses this narrative structure in taking the viewer on a voyage from the outer reaches of the universe down through the structure of an atom, then back to our world.

118 "Surface Tension" was originally published in *Galaxy Science Fiction*, August 1952. The novelette was later revised and merged with another story ("Sunken Universe" [1942]) as book 3 of Blish's fix-up novel about "pantropic" technologies, *The Seedling Stars* (1957). On the inclusion of "Surface Tension" among the top fifteen stories selected by SFWA for the Science Fiction Hall of Fame, see Silverberg, introduction to *The Science Fiction Hall of Fame*, xii. On Blish's place in pre-nanotechnological discourse, see Landon, "Less Is More," 132–33.

119 Blish, "Surface Tension," 405–6; hereafter cited in the text.

120 Lacan, "Mirror Stage," 77, 78, 75.

121 See Derrida, *Of Grammatology* and *Dissemination*.

122 Lacan, *The Seminar of Jacques Lacan, Book XI*, 45.

123 "The nanoManipulator," 1.

124 This jouissance would always remain unthinkable, however, inasmuch as it could only exist outside the symbolic; see Lacan, *The Seminar of Jacques Lacan, Book XX*.

125 The nanotechnological concept of "tunneling" here is very similar to what Timothy Morton has called "digging"; see Morton, "Wordsworth Digs the Lawn." Morton's notion of "digging" neither makes the real out to be an inaccessible remainder of thought, nor sublimes it out of touch as some inviolable or originary pure essence, but discovers the real through interacting with it, entering it with the tools one has "at hand" and digging it: "I use the term 'digging' in a precise (originally African-American, later countercultural) sense: to become perceptually involved in, to immerse oneself in, a phenomenological enrichment of 'to understand, to appreciate, like.' 'Digging' is a form of enjoyment—of labor as consumption, and, in reverse, of consuming as a form of production" (318). Digging and its enjoyment parallels nanovisual tunneling and its jouissance.

126 The story in fact opens with a discussion of hubris and rejects the concept itself as anthropocentric arrogance. Following the original crash landing, the pilot says:

If I were a religious man . . . I'd call this a plain case of divine vengeance. . . . It's as if we've been struck down for—is it *hubris*, arrogant pride? . . . It takes arrogant pride to think that you can scatter men, or at least things like men, all over the face of the Galaxy. (394)

But one of the scientist-colonists responds:

I suppose it does. . . . But we're only one of several hundred seed-ships in this limb of the Galaxy, so I doubt that the gods picked us out as special sinners. . . . If they had, maybe they'd have left us our ultraphone, so the Colonization Council could hear about our cropper. Besides . . . we try to produce men adapted to Earthlike planets, nothing more. We've sense enough—humility enough, if you like—to know we can't adapt men to Jupiter or to Tau Ceti. (395)

The drive to overcome the limits of the possible in this story is therefore not a question of metaphysical transgression, or transgression of physical realities and their material limitations, but a transgression of the limits imposed on human beings by their own inhibitions, their own imaginative failures, and their own metaphysics.

127 Lacan, "Mirror Stage," 78.

128 Ibid.

129 Sturgeon, "Microcosmic God," 96. The story was originally published in *Astounding Science Fiction*, April 1941. Like "Surface Tension," "Microcosmic God" anticipates many nanotech concerns, and it was likewise voted one of the fifteen most important stories from 1929 to 1964 by SFWA; see Silverberg, introduction to *The Science Fiction Hall of Fame*, xii. On the place of Sturgeon's story in the literary history of nano, see Landon, "Less Is More," 132–33. Sturgeon's story has been equally influential in another field of "science (fiction)" with close ties to nanotechnology: artificial life. See Helmreich, *Silicon Second Nature*, 88. Already in 1941, Sturgeon's story could thematize the intersection of nanotech's logic of control with our ethical responsibility toward the molecular other—an intersection made increasingly important for research in both nano and artificial life as new molecular modes of postbiological life become increasingly possible, as we will see in chapter 4.

130 Baxter, "The Logic Pool"; hereafter cited in the text. "The Logic Pool" was originally published in *Asimov's Science Fiction*, June 1994.

131 Eigler, quoted in Frankel, "Capturing Quantum Corrals," 261.

132 Binnig and Rohrer, "In Touch with Atoms," S327.

133 Levinas, *Totality and Infinity*, 223.

3. THE HORRORS OF GOO

1 See Kulinowski, "Nanotechnology: From 'Wow' to 'Yuck'?" Kulinowski documents the increase of nanofears relative to nanohopes—the growth of a "yuck

index" relative to a "wow index"—and compares her findings to the shifting of wow/yuck indices in other fields like genetic engineering and informatics, suggesting a common pattern of social response to the development of new technosciences. In a similar vein, Toumey has mapped the ecology of cultural narratives about nanotechnology and shows that the rise of mythologies of "extreme nanophobia" in response to mythologies of "extreme nanophilic hyperbole" serves to effectively erase more moderate or nuanced nanonarratives from public discourse; see Toumey, "Narratives for Nanotech." See also Laurent and Petit, "Nanosciences and Its Convergence with Other Technologies." On nanophobia in recent science fiction, see Dinello, *Technophobia*, 223–45.

2 Drexler, Peterson, and Pergamit, *Unbounding the Future*, 278.

3 Cautionary accounts of nano have grown legion. For some prominent examples, see Altmann, "Military Uses of Nanotechnology"; Altmann and Gebrud, "Military, Arms Control, and Security Aspects of Nanotechnology"; Borm et al., "Potential Risks of Nanomaterials"; Bostrom, "Existential Risks"; Mehta, "Some Thoughts on the Economic Impacts of Assembler Era Nanotechnology"; Moor and Weckert, "Nanoethics"; Oberdörster, "Manufactured Nanomaterials"; Posner, *Catastrophe*; Preston, "The Promise and Threat of Nanotechnology"; HRH the Prince of Wales, "Menace in the Minutiae"; Rees, *Our Final Hour*; Roberts, "Deciding the Future of Nanotechnologies"; Robison, "Nano-Ethics"; Seaton and Donaldson, "Nanoscience, Nanotoxicology, and the Need to Think Small"; and Sten, *Souls, Slavery, and Survival*. For perspectives on the ethical, social, and physical risks of nanotechnology offered by several practicing nanoscientists, see Berne, *Nanotalk*.

4 For examples, see Arnall, *Future Technologies, Today's Choices*; Center for Responsible Nanotechnology, "Dangers of Molecular Manufacturing"; Freitas, "Molecular Manufacturing"; Jacobstein, "Foresight Guidelines" (Version 6.0); Phoenix and Drexler, "Safe Exponential Manufacturing"; Royal Society and the Royal Academy of Engineering, *Nanoscience and Nanotechnologies*; and Swiss Re, *Nanotechnology: Small Matter, Many Unknowns*.

5 See ETC Group, *The Big Down* and "Size Matters!" The Canadian-based ETC Group (Action Group on Erosion, Technology, and Concentration), a technology-environmental watchdog organization, publishes regularly on the anticipated dangers of the nanotechnology age, focusing especially on risks involved in the convergence of nanotech with other sciences like biotechnology, information science, and cognitive science. See http://www.etcgroup.org. On the environmental debates surrounding nanoscience, see Schwarz, "Shrinking the Ecological Footprint."

6 McKibben, *Enough*, 119.

7 Joy, "Why the Future Doesn't Need Us," 244.

8 Marshall, "Future Present," 155.

9 Whitesides, "Self-Assembling Materials," 146–47. See also Whitesides, Mathias, and Seto, "Molecular Self-Assembly and Nanochemistry."

10 See Winner, *Autonomous Technology*; and Kelly, *Out of Control*.

11 Heidegger, "The Question Concerning Technology," 16.

12 Western cultural perceptions of autonomous technology as a monstrous inversion of natural order can be traced at least to the early modern period; see Huet, *Monstrous Imagination*; and Hanafi, *The Monster in the Machine*. As Huet and Hanafi document, historical views of monstrosity, whether biological or technological, often hinged on perceived violations of "natural law" and the "illegitimate" disordering of metaphysical binaries (e.g., man over woman, form over matter, father over progeny, master over slave). See also Davidson, "The Horror of Monsters"; and Todd, *Imagining Monsters*.

13 See Hansson, "Great Uncertainty about Small Things"; and Dupuy and Grinbaum, "Living with Uncertainty."

14 Freitas, "Some Limits to Global Ecophagy by Biovorous Nanoreplicators."

15 Drexler, *Engines of Creation*, 172–73.

16 Smalley, "Nanotechnology, Education, and the Fear of Nanobots," 145–46.

17 Bueno, "Drexler-Smalley Debate"; cf. Bensaude-Vincent, "Two Cultures of Nanotechnology?"

18 National Nanotechnology Initiative, "Frequently Asked Questions" (2003 update). The NNI has since removed these statements from its FAQ page.

19 Roco, "Interview with Dr. Mihail Roco."

20 Mehta, "The Future of Nanomedicine Looks Promising, But Only if We Learn from the Past." See also Rip, "Folk Theories of Nanotechnologists."

21 Schuler, "Perception of Risk and Nanotechnology," 283.

22 B-M Information, "Nanotechnology in Need of Successful Communication," 1.

23 While the NNI reserves a portion of its funding budget for social, ethical, and risk analyses, its early years have suggested a worrying inability to fruitfully employ the majority of these funds; see Mnyusiwalla, Daar, and Singer, "Mind the Gap." These researchers criticize the NNI and the global nanocommunity for failing to adequately address ethical and risk issues, suggesting that unless nanotechnology begins policing itself, a moratorium may be its only future: "The call by ETC for a moratorium on deployment of nanomaterials should be a wake-up call for NT [nanotechnology]. The only way to avoid such a moratorium is to immediately close the gap between the science and ethics of NT. The lessons of genomics and biotechnology make this feasible. Either the ethics of NT will catch up, or the science will slow down" (R12).
    At the same time, the efforts the NNI has made to directly address social and ethical implications of nano—as reported, for example, in the essay collections edited by Roco and Bainbridge, *Societal Implications of Nanoscience* and *Nanotechnology: Societal Implications*—often appear to sever the "ethical" from the "technical," thereby shielding the "science itself" from critique; see Lewenstein, "What Counts as a 'Social and Ethical Issue' in Nanotechnology?" Likewise, among various social groups, concerns about risk and ethics may be very different from those currently acknowledged by the NNI; see Schummer, "Soci-

etal and Ethical Implications of Nanotechnology." From a more cynical perspective, the NNI's publicized goal to undertake socioethical impact studies may simply be a strategy to attenuate popular anxieties and curtail critical backlash while enabling nanoresearch to proceed unhindered; see Berube, *Nano-Hype*, 305–34. Similarly, López argues that the science fiction features of nanoscience themselves strategically obscure viable risks by closing temporal gaps so tightly that the utopian nanofuture would seem to have already arrived, leaving no space for perceptions of danger or ethical considerations; see López, "Bridging the Gaps" and "Compiling the Ethical, Legal and Social Implications of Nanotechnology."

24 Modzelewski, quoted in Pethokoukis, "Government Says 'No' to Federally Funded Nanobots."

25 Modzelewski, "Industry Can Help Groundbreaking Nanotech Bill Fulfill Its Promise."

26 Smalley, "Smalley Concludes," 42.

27 National Science and Technology Council, *National Nanotechnology Initiative: Leading to the Next Industrial Revolution*. On the economic aspirations driving nanoscience policy, see A. Johnson, "The End of Pure Science."

28 Commentators have often characterized Drexler as "nanotechnology's unhappy father," not only because his "child" seems to be turning out quite differently than he had hoped (the NNI seemingly wants nanotech to be more like industrial chemistry and less like molecular assemblers), but also because some sectors of the nanoscape (e.g., the self-fashioned "legitimate scientific community") have ostracized him, challenging his claims for paternity. Adopting nano as their own, these stepfathers insist that Drexler fatally misunderstands the child, that he spreads fearful rumors about the child's association with hideous nanobots, that he is, in short, a bad father; see "Nanotechnology's Unhappy Father," 41–42.

29 See Jacobstein, "Foresight Guidelines," and Center for Responsible Nanotechnology, "Gray Goo Is a Small Issue."

30 Phoenix and Drexler, "Safe Exponential Manufacturing," 871.

31 Phoenix, "Don't Let Crichton's *Prey* Scare You." Several prominent scientists have denounced the technical accuracy of Crichton's fictive nanobot swarms; for example, see Dyson, "The Future Needs Us!" On "the *Prey* effect" in shaping public perceptions of nano, see Cobb and Macoubrie, "Public Perceptions about Nanotechnology." On the status of *Prey* in the discourse of nanotechnology, see Bowman, Hodge, and Binks, "Are We Really the Prey?"

32 Shelley, *Frankenstein*, 34.

33 Kristeva, *Powers of Horror*, 3.

34 Derrida, *Dissemination*, 268n67.

35 Kristeva, *Powers of Horror*, 4.

36 Ball, "The Robot Within: The Tools to Create an Army of Replicating Nanobots May Lie in Your Own Cells," 50.

37 Ian Gibson, quoted in Radford, "Brave New World or Miniature Menace?" 3.

38 Kristeva, *Powers of Horror*, 53.

39 Binnig and Rohrer, "Scanning Tunneling Microscopy," 615.

40 Ibid.

41 Ibid.

42 The "size matters" pun is ubiquitous in nanodiscourse. For just a few examples, see ETC Group, "Size Matters!"; Akin, "Nanotechnology: SIZE MATTERS"; and Ball, "When Size Does Matter."

43 Theis, "Nanotechnology: A Revolution in the Making," 11.

44 Waddington, "Grußwort/Preface," 4.

45 Uldrich and Newberry, *The Next Big Thing Is Really Small*.

46 "Nothing 'Nano' about It."

47 Harrow, "Of Numbers and Nature: Size Matters."

48 European Commission, Directorate-General for Research, *Nanotechnology: Small Science with a Big Potential*.

49 Saulnier, "Size Matters (Smaller Is Better)," 43.

50 Sargent, *The Dance of Molecules*, 210. The nanoblogger Howard Lovy has recently addressed such inflationary nanorhetoric in his blog entry "The Great Little Balls of Britain," mocking the virile posturing of the engineers responsible for the giant buckyball crowning the Bristol Centre for Nanoscience and Quantum Information: "Yes, yes, men. You are correct. Buckminsterfullerenes are strong, and so are you. But I hate to say this, guys, since you seem so . . . um . . . confident. But in real life, those buckyballs are really, really . . . yeah, really . . . tiny. Still, you blokes should go have a pint in celebration of your . . . erection . . . on the tip of the new Bristol (UK) Centre for Nanoscience and Quantum Information" (ellipses in original).

51 Drexler, quoted in "Foresight Update 53."

52 On the centrality of blindness or prohibition in structures of patriarchal authority, see Lacan, "Seminar on 'The Purloined Letter'"; and Lévi-Strauss, *The Elementary Structures of Kinship*.

53 A compilation of position statements by Drexler and other nanotheorists about the absence of molecular manufacturing in the $3.7 billion U.S. Twenty-first Century Nanotechnology Research and Development Act (2003) can be found in Rawstern, "Omission in the 21st Century Nanotechnology Research and Development Act." The act did make allowance for "a one-time study to determine the technical feasibility of molecular self-assembly for the manufacture of materials and devices at the molecular scale" (United States Congress, Senate, Committee on Commerce, Science, and Transportation, *21st Century Nanotechnology Research and Development Act*). But this textual trace would seemingly mark the last gasp of a smothered question—whose "one-timeness" suggests an order of "Do Not Resuscitate"—thus making clear the sociological actions of blindness here. For of course "molecular self-assembly" is "feasible" (organisms are particularly good examples of such feasibility). What the omission does is

exclude whole nanoresearch programs from major funding access: namely, those programs associated with the specter of "molecular autonomy" (i.e., concepts of "self-action," "self-assembly," "self-replication," even "self-awareness") and, accordingly, the more ominous specter of gray goo. Reviewing the NNI in 2005, a National Research Council committee carried out the "one-time study" by convening nanoscience experts (including Drexler, Eigler, Merkle, and Montemagno) to discuss molecular self-assembly and nanomanufacturing. The committee's report, *A Matter of Size*, concluded:

> The committee found the evaluation of the feasibility of these ideas to be difficult because of the lack of experimental demonstrations of many of the key underlying concepts. . . . Although theoretical calculations can be made today, the eventually attainable range of chemical reaction cycles, error rates, speed of operation, and thermodynamic efficiencies of such bottom-up manufacturing systems cannot be reliably predicted at this time. Thus, the eventually attainable perfection and complexity of manufactured products, while they can be calculated in theory, cannot be predicted with confidence. Finally, the optimum research paths that might lead to systems which greatly exceed the thermodynamic efficiencies and other capabilities of biological systems cannot be reliably predicted at this time. Research funding that is based on the ability of investigators to produce experimental demonstrations that link to abstract models and guide long-term vision is most appropriate to achieve this goal. (107–8).

Although open to the "eventually attainable perfection" and future benefits of molecular nanotechnology, the committee closed its investigation by suggesting that the NNI allocate research funding only in consequence of "experimental demonstrations" rather than in support of simulations or abstract models, clarifying that real science should "guide long-term vision" and not vice versa. This recommendation, of course, is commensurate with other ongoing efforts that try to dissociate nanotechnology from science fiction while simultaneously espousing science fiction goals.

54 On the personal and bombastic accusations of blindness exchanged between Drexler and his supporters on the one hand, and Roco, Smalley, and their supporters on the other, see Berube and Shipman, "Denialism."

55 Ober, quoted in Saulnier, "Size Matters," 46.

56 Drexler, quoted in Lovy, "Drexler on 'Drexlerians.'" Drexler has also claimed that anonymous forces of the NNI attempted, unsuccessfully, to uninvite him from the "Imaging and Imagining Nanoscience" conference at the University of South Carolina in 2003 because of his outspoken views on the dangers of nanotechnology, especially military applications; see Lovy, "National Nanotechnology 'Disarmament' Initiative." Berube and Shipman clarify this incident: "For the record, Mihail Roco, who chairs the President's Council of Advisors on Science and Technology, questioned our decision to invite Drexler to speak

at the conference ["Imaging and Imagining Nanoscience"], something which Drexler alluded to several times during his presentation. Although Roco made no effort to pressure us into disinviting Drexler, another speaker did refuse to speak on the same dais with him [Drexler] and she chose not to attend" ("Denialism," 22).

57 Drexler, "Nanotechnology: From Feynman to Funding," 21.

58 Hans Glimell argues that the "gatekeeping activities" of the legitimate nanotech community (arising largely in reaction to Bill Joy's article "Why the Future Doesn't Need Us") demanded the banishment of Drexler and even Feynman because of the extent to which they both could be associated with perceived dangers of nanomachines. But the banishment of Drexler, in Glimell's view, could ultimately prove problematic for the NNI. In choosing a strategy of simply sweeping problems under the bed to counteract the alternative strategy of relinquishment represented by Joy, they lose out on the "third option" represented by Drexler: visionary utopianism coupled with prudent caution about nanotechnology. (Moreover, they miss out on the grand Feynman vision and all its appealing features, which were necessary for generating excitement about nanotechnology in the first place.) See Glimell, "Grand Visions and Lilliput Politics." See also McCray, "Will Small Be Beautiful?"

59 Drexler, "Nanotechnology: From Feynman to Funding," 25.

60 Ibid.

61 Žižek, *Enjoy Your Symptom*, 89–90.

62 Ibid., 90.

63 Ibid., 73–74.

64 Ball, "2003: Nanotechnology in the Firing Line."

65 Lovy, "Welcome to Nano Reality TV, Where the Show Is Mistaken for Truth."

66 Boehlert, "Opening Statement for Hearing on Nano Consequences."

67 Tolles, "National Security Aspects of Nanotechnology," 222; hereafter cited in the text.

68 This discursive construction of irrationality or madness depends on the epistemic field of classical humanism and its concomitant metaphysics of the rational and the enlightened; see Foucault, *History of Madness*. On psychosis as a refusal of lack (castration) in the register of the humanistic symbolic, see Lacan, *The Seminar of Jacques Lacan, Book III*, 201.

69 Kroto, quoted in Henderson, "Prince's 'Goo' Fears Are Nonsense, Say Experts." On the question of "legitimate" expertise in nanoscience and the authoritarian rhetoric of "rational deference," see Sanchez, "The Expert's Role in Nanoscience and Technology."

70 On designations of "hard science" as gendering natural knowledge, see Keller, *Reflections on Gender and Science*, 75–94. The concept of "hard science," as Keller shows, is one manifestation of a masculinist principle active in technoscience (i.e., the technoscientific phallus) that ensures "the predisposition to kinds of explanation that posit a single central governor" (155). Similarly, Brian Attebery

has shown that fictive representations of science typically gender according to the presumed "hardness" or "softness" of the scientific field they depict; hence "hard science fiction" has typically been understood as a masculine form of the genre, while "soft science fiction" has been understood as feminine; see Attebery, *Decoding Gender in Science Fiction*, 46–61.

71 "Editorial: Growing Nanoknowledge," A14. The editorial concludes that, because nanotechnology is "hard science" supported by many credentialed and responsible researchers, there is no cause for fear (and even less cause for a moratorium): "Scientists and businessmen alike should pursue nanoknowledge."

72 Laqueur, *Making Sex*.

73 Hansen, *Embodying Technesis*.

74 Butler, *Bodies That Matter*, 7–8. Butler has extensively argued that the symbolic economy of humanism is fundamentally structured by binary gender: "The mark of gender appears to 'qualify' bodies as human bodies; the moment in which an infant becomes humanized is when the question, 'is it a boy or girl?' is answered. Those bodily figures who do not fit into either gender fall outside the human, indeed, constitute the domain of the dehumanized and the abject against which the human itself is constituted" (Butler, *Gender Trouble*, 142). Such humanist gender encoding has sometimes introduced anthropocentric bias in the scientific study of nonhuman domains: for examples at the level of organisms, see Haraway, *Primate Visions*; at the level of cells, see Martin, "The Egg and the Sperm"; at the level of biomolecules, see Spanier, *Im/Partial Science*; at the level of physical systems, see Hayles, "Gender Encoding in Fluid Mechanics."

75 Butler, *Bodies That Matter*, 27–55, 31.

76 Irigaray, *This Sex Which Is Not One*, 110. On the metaphysical history of matter and its purported femininity as a back-formation of masculinist thought, from classical philosophy through psychoanalysis, see Irigaray, *Speculum of the Other Woman*. Hayles extends Irigaray's assertions on the gendering of fluid mechanics relative to solid mechanics and offers a more robust historical account in "Gender Encoding in Fluid Mechanics."

77 Lacan, "Signification of the Phallus," 581.

78 Lacan, "Seminar on 'The Purloined Letter,'" 30. However, Derrida has argued that the phallic itinerary analyzed by Lacan is always threatened by its own potential for dissemination; in other words, "a letter does *not always* arrive at its destination, and from the moment that this possibility belongs to its structure one can say that it never truly arrives, that when it does arrive its capacity not to arrive torments it with an internal drifting" (*The Post Card*, 489). Both Lacan and Derrida here seem to be getting at the same point in the end, namely, that the humanist symbolic field functions only to the extent that it excludes or blinds itself to the perpetual risk of disseminating matters, which it therefore does not exclude; see B. Johnson, "The Frame of Reference."

79 Hanafi, *The Monster in the Machine*, 93. For further discussion of monstrosity

as a deconstructive disordering of metaphysical boundaries, see Haraway, "The Promises of Monsters"; and C. Milburn, "Monsters in Eden."

80 On theoretical paradigms of self-capacitating matter as posing a revolutionary challenge to authoritarian systems of control—political, scientific, or otherwise—see Merchant, *The Death of Nature*; and Rogers, *The Matter of Revolution*.

81 Creed, *The Monstrous-Feminine*. According to Creed, the monstrous-feminine is the symbolic "negative force" in cultural production that bodies forth horrific images of the "voracious maw, the mysterious black hole that signifies female genitalia which threatens to give birth to equally horrific offspring as well as threatening to incorporate everything in its path. This is the generative archaic mother, constructed within patriarchal ideology, as the primeval 'black hole,' the originating womb which gives birth to all life" (27).

82 See both volumes of Theweleit's *Male Fantasies*.

83 Alex Travis, a nanobiologist at Cornell University, quoted in Staedter, "Sperm Power: New Tool for Nanobots."

84 McDonald, *Evolution's Shore*, 80; hereafter cited in the text. This novel was also published in the United Kingdom as *Chaga*.

85 Crichton, *Prey*, 336; hereafter cited in the text.

86 See Goldberg, "Recalling Totalities."

87 This salvific family drama, then, is entirely directed by men, and the woman becomes a mere conduit for male autogenesis; see Goscilo, "Deconstructing *The Terminator*."

88 See Penley, "Time Travel, Primal Scene and the Critical Dystopia." Penley analyzes the "primal scene" in *The Terminator* as enacting a fantasy of incestuous Oedipal desire. Though as Goldberg suggests in "Recalling Totalities," *The Terminator*'s primal time-travel scene also fulfills other "forbidden" desires: sex with oneself, or sex between men—a desire of the *phallus for itself* that Goldberg detects throughout the filmography of Arnold Schwarzenegger, such that Arnold becomes the very object of desire the anthropic regime seeks to possess, and to love.

89 Several critics have analyzed the feminine or feminizing qualities of the liquidic T-1000 as compared to the hard masculinity of the original Terminator (T-800, a.k.a. T-101); see, for example, Springer, *Electronic Eros*, 95–124; Dery, *Escape Velocity*, 260–70; and Bukatman, *Terminal Identity*, 303–11.

90 Cyborgs, even when they subvert gender constructions, seem to always end up redeploying gender by the very fact that the "skin" or visual surface of cyborg bodies enacts a history of sexualized images; see Kakoudaki, "Pinup and Cyborg." While the redeployment of gender may offer novel formations, gender is nevertheless always there, its history present at the surface, even if contested. The "techno-body" is a surface whose reworking or rewiring usually entails a reinscription of gender traditions, as well as racial and class categories,

that, even in being made malleable, remain visible; see Balsamo, *Technologies of the Gendered Body*.

91 Goldberg, "Recalling Totalities," 245–48.

92 Riviere, "Womanliness as a Masquerade."

93 Lovejoy, "Re: Miscellaneous."

94 Freitas, "Police Nanites and Nanorobot Population Limits."

95 Lovejoy, "Re: Miscellaneous."

96 On the role of abortion in *The Terminator* as another way in which the film seeks to erase women in the service of androcentric autogenesis, see Goscilo, "Deconstructing *The Terminator*." But at the same time, the question of abortion developed in the films, especially in *T2*, explodes the humanist binaries that animate the so-called abortion debate in modern America. In the abortive actions of the *Terminator* universe, it becomes impossible to make distinctions at the level of bodies or persons, and the question must instead be thought in terms of relations, networks, and temporality; see Mason, "Terminating Bodies." This cyborgism of abortion in the *Terminator* films—perhaps best signified in *T3* by the electronic switching between "abort" and "terminate" as supplementary computer commands that both reinforce and cancel each other in displaying across the visual image of John Connor—opens escape routes from deadlocked gender binaries and singularities of "personhood," inviting instead multiplicities and processes of becoming-posthuman.

97 The Terminator's assertion that "desire is irrelevant" at this point of having been (re)absorbed into the machinic regime and his subsequent violent revolt suggest a hysterical response to the crisis or end—indeed, the utter "irrelevance"—of "unified white masculine subjectivity" which is animated by cyborg fictions generally; see Fuchs, "Death Is Irrelevant," 282.

98 Arnold Schwarzenegger, in "Commentary by Actors Arnold Schwarzenegger, Nick Stahl, Claire Danes, Kristanna Loken, and Director Jonathan Mostow," DVD audio track 3, *Terminator 3: Rise of the Machines*, 2-disc widescreen edition DVD (Burbank: Warner Home Video, 2004).

99 Gray, *Cyborg Citizen*, 195.

100 On the rhetoric of security in discussions of military nanotechnology relative to constructions of soldiers' and civilians' bodies, see Gray, *Cyborg Citizen*, 55–65; Parisi and Goodman, "The Affect of Nanoterror"; and C. Milburn, "Nanowarriors."

101 For coverage of current military nanotech R&D, see Altmann, *Military Nanotechnology*.

102 Ratner and Ratner, *Nanotechnology and Homeland Security*; hereafter cited in the text. The Ratners are also the authors of a previous highly successful "introduction" to nanotechnology, *Nanotechnology: A Gentle Introduction*. This text features much of the same rhetoric we will see in the following discussion, especially the insistence that nanotech has nothing to do with molecular self-

organization or self-replication—that is, it has nothing to do with nanobots. On ways in which the Ratners' texts, amid a horde of popular nanobooks, contribute to popular understanding, or "misunderstanding," of what nanotechnology is or professes to be, see Schummer, "Reading Nano."

103 Theweleit, *Male Fantasies: Volume 1*, 244.

104 On the gendered history of metaphysical conceptions of mind and body, scientific subject and material object, see Jordanova, *Sexual Visions*; Park, *Secrets of Women*; Schiebinger, *The Mind Has No Sex?*; and Spelman, "Woman as Body."

105 On GE's futurist orientation under Immelt, see Reiss, "Size Matters."

106 Blohm, quoted in Teresko, "The Next Material World," 46.

107 Blohm, quoted in Baker and Aston, "The Business of Nanotech," 66.

108 Lacan, "Seminar on 'The Purloined Letter,'" 10, 17; Lacan, "Signification of the Phallus," 582–83.

109 Fictive representations of science regularly figure the technologically enhanced male eye as the source of inseminatory power over both women and the universe; as Attebery writes: "The act of seeing can lead not only to (symbolic) sexual release but still further, to impregnating the universe. . . . Because the scientific gaze is so insistently masculine, whatever it touches upon is feminized" (*Decoding Gender in Science Fiction*, 52).

110 Mulvey, "Visual Pleasure and Narrative Cinema." Mulvey later said that "Visual Pleasure" was written in 1973 "polemically and without regard for context or nuances of argument," but instead for the intellectual and political needs of the historical moment (Mulvey, *Visual and Other Pleasures*, vii). Has that moment so fully passed, after all? Ads like "Beauty and Brains" might suggest otherwise. Nevertheless, more recent efforts to study the gender dynamics of cinematic "love stories" have complicated and ambiguated masculine and feminine modes of the gaze and the montage; see, for example, MacKinnon, *Love, Tears and the Male Spectator*.

111 Kristeva, *Powers of Horror*, 204.

112 Ibid., 206.

113 On the "informe," or formless, see Bataille, *Visions of Excess*; and Bois and Krauss, *Formless*. On corporeal volatility, see Grosz, *Volatile Bodies*.

114 See B. Johnson, *A World of Difference*; Braidotti, *Metamorphoses*; and B. Clarke, *Posthuman Metamorphosis*.

4. NANO/SPLATTER

1 Roco, "Nanotechnology: Convergence with Modern Biology and Medicine" (2003). Nanodiscourse uses the terms "nanobiology," "nanobiotechnology," and "bionanotechnology" synonymously—a specific lexicon has yet to be stabilized—for what Roco, following Richard Smalley, defines as "wet nanotechnology," or "the field that applies the nanoscale principles and techniques to understand and transform biosystems (living or non-living) and which uses

biological principles and materials to create new devices and systems integrated from the nanoscale" (337). I will use "nanobiology" to designate "wet nanotechnology" generally, emphasizing the extent to which nanoscience, as it increasingly colonizes the experimental systems of biotechnology, transforms the foundations of what has traditionally been called "biology."

2 The epistemic shrinkage of the organism registered in the 1990s with the "discovery" of nanobacteria, seemingly self-replicating particles as small as 30 nm across, found everywhere from human kidney stones to ancient geological strata to Martian meteorites. Despite Jack Maniloff's demonstration that a cell must be at least 140 nm across to contain requisite DNA and proteins (Maniloff, "Nannobacteria"), "nanobacteriologists" maintain that nanobacteria are alive because they metabolize and reproduce themselves. Though still facing skepticism, results released by Mayo Clinic researchers in 2004 suggest that these creatures, smaller than viruses, are probably "unique living organisms" (Miller et al., "Evidence of Nanobacterial-Like Structures," H1115). Similarly, consider the 2002 announcement by the geneticists J. Craig Venter and Hamilton Smith of their intention to fabricate a "minimalist organism" and thereby identify "a molecular definition of life": the minimum number of biomolecules needed to sustain and reproduce a homeostatic entity (quoted in Gillis, "Scientists Planning to Make a New Form of Life," A1). Research on nanobacteria and "minimalist organisms" represents not only a scale change but a conceptual reorientation of the biological gaze toward the "vitality" of nanoscale systems.

3 Ostman, "Nanobiology." See also Ostman, "The Nanobiology Imperative."

4 Ostman, "Bioconvergence."

5 Ostman, "Nanobiology."

6 Addressing the future destination as already inevitable, the scene of disintegration operates through what Jacques Derrida calls the "post effect"; see Derrida, *The Post Card*. I will say more of this by the end. (Have I destined it?)

7 Ostman, "Bioconvergence."

8 Pethes, "Terminal Men," 169.

9 Kauffman, *Investigations*, 8, 72.

10 Shirley, *Crawlers*, 362; hereafter cited in the text.

11 Goonan, *Queen City Jazz*, 39.

12 Ostman, "Bioconvergence."

13 On the cultural history of artificial-life research, see Helmreich, *Silicon Second Nature*. On evolutionary robotics and autonomous-agent theory, see J. Johnston, "A Future for Autonomous Agents." On prebiotic molecular evolution, see Cairns-Smith, *Genetic Takeover and the Mineral Origins of Life*. On the way these sciences instantiate a postbiological and posthuman era, see Doyle, *On Beyond Living*; and Hayles, *How We Became Posthuman*.

14 S. Rasmussen et al., "Transitions from Nonliving to Living Matter," 965.

15 Ibid., 965.

16 Seeman, "Nanotechnology and the Double Helix," 65, 75, 75.

17 ETC Group, "Green Goo," 5. The ETC Group uses the term "green goo" (troping on "gray goo") to refer to the proliferation of nanobiomaterials in industrial, medical, agricultural, military, and information science applications, as well as the novel forms of machinic life enabled by the hybridization of nanotech and biotech: "Nanobiotechnology will create both living and non-living hybrids previously unknown on earth" (6).

18 Freitas and Merkle, *Kinematic Self-Replicating Machines*, 1. On the history of self-replicating machines dating back to Alan Turing, and the epistemic significance of this concept for the development of nanotechnology, see Bueno, "Von Neumann, Self-Reproduction, and the Constitution of Nanophenomena."

19 As Kevin Kelly points out, "Nanotechnology has the same potential for artificial [nonbiological] evolution as biological molecules" (*Out of Control*, 310). Kelly's enthusiastic account suggests that "biologized" technologies fundamentally escape human control in their unpredictable evolvability. Precisely for this reason, some nanotechnologists working on self-replicating molecular systems emphasize the need to stifle evolutionary tendencies: "Microscale reproducers ... may be able to generate offspring up to a million times faster [than macro-scale reproducers], thus could be far more likely to randomly yield productive modifications, hence to 'evolve.' ... Microscale replicators may therefore be viewed as inherently more risky from a public safety standpoint and so the need for adherence to design guidelines to forestall unplanned system evolution is more urgent in this realm" (Freitas and Merkle, *Kinematic Self-Replicating Machines*, 176).

20 Deleuze and Guattari, *A Thousand Plateaus*, 59–60.

21 On the history of these forms of biological science (generally speaking, molecular biology and genetics) that understand life as the controlled, coordinated product of a single master molecule—whether protein or amino acid—or a particular informational code-script, see Kay, *The Molecular Vision of Life* and *Who Wrote the Book of Life*; Keller, *The Century of the Gene*; and Doyle, *On Beyond Living*. The nanomachinic entity escapes these regimes of genetic control and molecular mastery—even (or perhaps especially) if it is itself made up of DNA—by being its own genetic source.

22 On homeostatic and autopoietic systems theories, see Hayles, *How We Became Posthuman*, 50–84, 131–60. Homeostasis and autopoiesis (the former from the work of Walter B. Cannon, the latter from Humberto Maturana and Francisco Varela) have had different scientific legacies, but both see biosystems as closed, self-referential operations that work autonomously to maintain boundaries. Yet as these systems are theoretically closed, they cannot easily accommodate evolution, symbiosis, or parasitic invasion; see Csányi and Kampis, "Autogenesis," which proposes a theory of autogenesis to account for "open" self-replicative biological systems. For a broad philosophical critique of self-referential, closed-systems models (revisioning living systems as sites of endless "viroid" contamination and rhizomatic change), see Ansell-Pearson, *Viroid Life*, 123–50. The

"opened" organism and the replicative machines of speculative nanobiology correspond more or less with these accounts.

23 Doyle, *On Beyond Living*. For a more expansive consideration of the postvital in contemporary culture, see Doyle, *Wetwares*.

24 Thacker, *Biomedia*, 28, 27 (see esp. 115–40 on nanomedicine's technical remediation of the postvital, postnatural body).

25 Deleuze and Guattari, *A Thousand Plateaus*, 409.

26 Lehn, "Toward Complex Matter," 4763, 4764. A leading supramolecular chemist, Lehn received the Nobel Prize in 1987 for his work on artificial proteins—retrospectively absorbed by nanotechnology discourse as a foundational step toward molecular manufacturing.

27 Deleuze and Guattari, *A Thousand Plateaus*, 411. For examples of nonorganic life made newly recognizable by recent scientific and philosophical paradigm shifts, see De Landa, "Nonorganic Life."

28 De Landa, *A Thousand Years of Nonlinear History*, 21.

29 Lehn, "Supramolecular Chemistry," 250.

30 Whitesides and Boncheva, "Beyond Molecules," 4769. See also Seeman and Belcher, "Emulating Biology."

31 Roco and Tomellini, eds., *Nanotechnology: Revolutionary Opportunities*, 18. For critiques of nanotech's logic of control, see Sarewitz and Woodhouse, "Small Is Powerful"; and Kearnes, "Chaos and Control."

32 Crandall, preface to *Nanotechnology: Research Perspectives*, vii–viii.

33 Sargent, *The Dance of Molecules*, xii.

34 Pickering, *The Mangle of Practice*.

35 Latour, *We Have Never Been Modern*, 142–45.

36 Lehn, "Supramolecular Chemistry," 251. For Lehn, contemporary nanoscience would appear, somewhat paradoxically, to have been produced by the "natural" nanotechnology of the machinic phylum in order to reflect upon itself. His inquiry approaches the reflexive logic of the "anthropic principle," on which see Bostrom, *Anthropic Bias*. Sandy Baldwin has written of nanotechnology's "discovery" of itself in the machinic phylum and analyzes the consequences for phenomenology in "Nanotechnology! (or SimLifeWorld)." On the way human history at large has danced and coevolved with the machinic phylum, see De Landa, *War in the Age of Intelligent Machines* and *A Thousand Years of Nonlinear History*.

37 Hayles, *How We Became Posthuman*, 288.

38 On the reflexive construction of the human and biological life in the modern era, see Foucault, *The Order of Things*: "When natural history becomes biology . . . in the profound upheaval of such an archaeological mutation, man appears in his ambiguous position as an object of knowledge and as a subject that knows" (312).

39 McCarty, *Splatter Movies*, 1. On literary splatter and its confrontational aesthetics of subversion and ideological revolution, see Sammon, "Outlaws."

40 McRoy, "There Are No Limits," 133, 130.

41 Halberstam, *Skin Shows*, 144. For more on the fragmentation of identity and gender in splatter films, see Clover, *Men, Women, and Chainsaws*.

42 Pulver, GURPS *Ultra-Tech 2*, 71.

43 Disintegration embodies a Deleuzian form of "perception [that] entails pulverizing the world, but also one of spiritualizing its dust" (Deleuze, *The Fold*, 87). Because disintegration exposes the organism as an autogenic assemblage of molecular agents—a dismantled "body without organs" (see *A Thousand Plateaus*, 149–66)—unrestricted (in principle) from the homeostatic closure of the cell, it departs from otherwise related perceptions, such as the autopoietic view of the organism as a collective of interdependent cellular entities, a "meshwork of selfless selves" (Varela, "Organism").

44 Marlow, *Nano*, 107; hereafter cited in the text.

45 Goodsell, *Bionanotechnology*, 32.

46 Seltzer, *Serial Killers*.

47 Kurzweil, *The Singularity Is Near*, 399.

48 Mano, Mao, and Heller, "A Miniature Biofuel Cell Operating in a Physiological Buffer," 12973. On potential applications of this device, see Ball, "Chemists Build Body Fluid Battery."

49 "Power from Blood Could Lead to 'Human Batteries.'"

50 J. Brown, "Vampire Bot."

51 Deleuze and Guattari, *Anti-Oedipus*, 286.

52 Foucault, *The Order of Things*, 387.

53 Goodsell, *Bionanotechnology*, 5.

54 Hall, "What I Want."

55 Bear, *Blood Music*, 130; hereafter cited in the text.

56 As the advertising slogan for Clive Barker's *Hellraiser* (1986), "There are no limits" encapsulates splatter's ethos of violent boundary destruction and its incessant aesthetic drive toward "the beyond"; see McRoy, "There Are No Limits," and Kern, "American 'Grand Guignol.'"

57 Compare Kauffman: "I want to say that autonomous agents are parts of the ontological furniture of the universe" because the coconstruction of the cosmos emerges "by the self-consistent search of autonomous agents for ways to make a living" (*Investigations*, 128, 243). Similarly, Lehn suggests that material self-organizing processes (the machinic phylum) express a common "drive to life," and that nanotechnology and supramolecular chemistry give access to "the spatial (structural) and temporal (dynamic) features of matter and . . . its complexification through self-organization, the drive to life" ("Toward Complex Matter," 4763).

58 Deleuze and Guattari, *Anti-Oedipus*, 286.

59 In *Beyond the Pleasure Principle*, Freud famously intuited a postbiological beyond as that which interferes with the circulation of the pleasure principle. He called it the death drive. But in understanding "death" as the only alternative

to the pleasures of life, Freud chose humanistic fatality over the possibility that the postbiological trace inhabiting the pleasure principle could instead be a circuitous pathway to "more life," a drive to molecularity. This is not to say that Freud's death drive is simply what I am describing as the postbiological, but rather that Freud's glimpse beyond the psycho-structural limits of life is analogous to the glimpse of machinic alterity offered by nanofiction.

60 Deleuze and Guattari, *Anti-Oedipus*, 285.

61 Ibid., 26.

62 J. Johnston, *Information Multiplicity*, 1–58.

63 Bear, "Introduction to 'Blood Music,'" 15.

64 That disintegration can act as a posthuman expression of love is suggested by the Nobel Prize–winning biochemist Arthur Kornberg. His book *For the Love of Enzymes* chronicles his heroic efforts of grinding and pulverizing inordinate numbers of organisms to encounter the "beloved" enzymes within. While my own organismic sensibilities cry out for these long-since-disintegrated organisms, I cannot help recognizing these laboratory scenes of disintegration as symptoms of a postbiological love for the molecular future that has been emerging since the early twentieth century.

65 Derrida, "Deconstruction and the Other," 123–24.

66 Derrida, *Points . . .* , 83.

67 Weinstone, *Avatar Bodies*, 41, 156.

68 Deleuze and Guattari, *A Thousand Plateaus*, 30.

69 Derrida, *The Post Card.*

70 Ostman, "Evolution into the Next Millennium." Such destining once again marks the humanist metaphysics that exists in deconstructive relationship to the posthumanism of nanotechnology. Hayles points to the recalcitrant humanism inhabiting transcendental nanofiction; see her analysis of *Blood Music* in *How We Became Posthuman*, 252–56. I would suggest, however, that this deconstructive dynamic between humanism and posthumanism (or between biology and postbiology) is what enables speculation on the "post" at all.

71 Attebery, "Dust, Lust, and Other Messages from the Quantum Wonderland," 161–69.

72 Nagata, *Limit of Vision*, 212, 326, 15.

BIBLIOGRAPHY

Akin, Jim. "Nanotechnology: SIZE MATTERS." *PC Magazine* 23, no. 12 (2004): 134–38.

Alexander, Brian. *Rapture: How Biotech Became the New Religion*. New York: Basic Books, 2003.

Alien. Directed by Ridley Scott. Screenplay by Dan O'Bannon. Twentieth Century–Fox, 1979.

Aliens. Directed by James Cameron. Screenplay by James Cameron. Twentieth Century–Fox, 1986.

Altmann, Jürgen. *Military Nanotechnology: Potential Applications and Preventative Arms Control*. London: Routledge, 2006.

———. "Military Uses of Nanotechnology: Perspectives and Concerns." *Security Dialogue* 35 (2004): 61–79.

Altmann, Jürgen, and Mark A. Gubrud. "Military, Arms Control, and Security Aspects of Nanotechnology." In *Discovering the Nanoscale*, ed. Davis Baird, Alfred Nordmann, and Joachim Schummer, 269–77. Amsterdam: IOS Press, 2004.

Anderson, Ben, Matthew Kearnes, and Robert Doubleday. "Geographies of Nanotechnoscience." Special section of *Area* 39 (2007): 139–75.

Ansell-Pearson, Keith. *Viroid Life: Perspectives on Nietzsche and the Transhuman Condition*. New York: Routledge, 1997.

Appenzeller, Tim. "The Man Who Dared to Think Small." *Science* 254 (1991): 1300.

Arnall, Alexander Huw. *Future Technologies, Today's Choices: Nanotechnology, Artificial Intelligence and Robotics; A Technical, Political, and Institutional Map of Emerging Technologies—A Report for the Greenpeace Environmental Trust*. London: Greenpeace Environmental Trust, 2003.

Asfari, Zouhair, and Jacques Vicens. "Molecular Machines." *Journal of Inclusion Phenomena and Macrocyclic Chemistry* 36 (2000): 103–118.

Ashley, Steven. "Nanobot Construction Crews." *Scientific American* 285, no. 3 (2001): 84–85.

Atkinson, William Illsey. *Nanocosm: Nanotechnology and the Big Changes Coming from the Inconceivably Small*. New York: AMACOM/American Management Association, 2003.

Attebery, Brian. *Decoding Gender in Science Fiction*. New York: Routledge, 2002.

———. "Dust, Lust, and Other Messages from the Quantum Wonderland." In *Nanoculture: Implications of the New Technoscience*, ed. N. Katherine Hayles, 161–69. Bristol: Intellect Books, 2004.

Auerbach, Erich. *Mimesis: The Representation of Reality in Western Literature*. 1946. Trans. Willard B. Trask. Fiftieth-anniversary edition. Princeton: Princeton University Press, 2003.

Avery, Todd. "Nanoscience and Literature: Bridging the Two Cultures." *New Solutions* 15 (2005): 289–307.

Badmington, Neil. *Alien Chic: Posthumanism and the Other Within*. London: Routledge, 2004.

Bainbridge, William Sims. *Across the Secular Abyss: From Faith to Wisdom*. Lanham: Lexington Books, 2007.

———. "Cyberimmortality: Science, Religion, and the Battle to Save Our Souls." *The Futurist* 40, no. 2 (2006): 25–29.

———. *Dimensions of Science Fiction*. Cambridge: Harvard University Press, 1986.

———. "The Impact of Science Fiction on Attitudes toward Technology." In *Science Fiction and Space Futures*, ed. Eugene M. Emme, 121–35. San Diego: American Astronautical Society, 1982.

———. "Massive Questionnaires for Personality Capture." *Social Science Computer Review* 21 (2003): 267–80.

———. *Nanoconvergence: The Unity of Nanoscience, Biotechnology, Information Technology and Cognitive Science*. Upper Saddle River, N.J.: Prentice Hall, 2007.

———. "Progress toward Cyberimmortality." In *The Scientific Conquest of Death: Essays on Infinite Lifespans*, ed. Immortality Institute, 107–22. Buenos Aires: Libros en Red, 2004.

———. "A Question of Immortality." *Analog* 122, no. 5 (May 2002): 40–49.

———. "Religions for a Galactic Civilization." In *Science Fiction and Space Futures*, ed. Eugene M. Emme, 187–201. San Diego: American Astronautical Society, 1982.

———. "Sociocultural Meanings of Nanotechnology: Research Methodologies." *Journal of Nanoparticle Research* 6 (2004): 285–99.

———. *The Spaceflight Revolution*. New York: Wiley Interscience, 1976.

Baird, Davis. *Thing Knowledge: A Philosophy of Scientific Instruments*. Berkeley: University of California Press, 2004.

Baird, Davis, Alfred Nordmann, and Joachim Schummer, eds. *Discovering the Nanoscale*. Amsterdam: IOS Press, 2004.

Baird, Davis, and Ashley Shew. "Probing the History of Scanning Tunneling Microscopy." In *Discovering the Nanoscale*, ed. Davis Baird, Alfred Nordmann, and Joachim Schummer, 145–56. Amsterdam: IOS Press, 2004.

Baker, Stephen, and Adam Aston. "The Business of Nanotech." *Business Week* 3920 (February 2005): 64–71.

Baldwin, Sandy. "Nanotechnology! (or SimLifeWorld)." *Culture Machine* 3 (2001). http://culturemachine.tees.ac.uk (web pages on file with author).

Ball, Philip. "Chemists Build Body Fluid Battery: Biofuel Cell Runs on Metabolic Energy to Power Medical Implants." *Nature News*, November 12, 2002. http://www.nature.com/news (visited May 13, 2007; web pages on file with author).

———. "It's a Small World." *Chemistry World* 1, no. 2 (2004): 30–37.

———. "The Robot Within: The Tools to Create an Army of Replicating Nanobots May Lie in Your Own Cells." *New Scientist* 177, no. 2386 (2003): 50–51.

———. "2003: Nanotechnology in the Firing Line." *nanotechweb.org*, December 23, 2003. http://nanotechweb.org (visited April 21, 2005; web pages on file with author).

———. "When Size Does Matter." *Nature* 349 (1991): 101–2.

Ballard, J. G. *Crash*. London: Jonathan Cape, 1973.

Balsamo, Anne. *Technologies of the Gendered Body: Reading Cyborg Women*. Durham: Duke University Press, 1996.

Barr, Marleen S. *Future Females, the Next Generation: New Voices and Velocities in Feminist Science Fiction Criticism*. Lanham, Md.: Rowman and Littlefield, 2000.

Bataille, Georges. *Visions of Excess: Selected Writings, 1927–1939*. Trans. Allan Stoekl with Carl R. Lovitt and Donald M. Leslie Jr. Ed. Allan Stoekl. Minneapolis: University of Minnesota Press, 1985.

Baudrillard, Jean. *Écran total*. Paris: Éditions Galilée, 1997.

———. *The Illusion of the End*. Trans. Chris Turner. Stanford: Stanford University Press, 1994.

———. "The Precession of Simulacra." In *Simulacra and Simulation*, trans. Sheila Faria Glaser, 1–42. Ann Arbor: University of Michigan Press, 1994.

———. "Simulacra and Science Fiction." In *Simulacra and Simulation*, trans. Sheila Faria Glaser, 121–27. Ann Arbor: University of Michigan Press, 1994.

———. *Simulacra and Simulation*. Trans. Sheila Faria Glaser. Ann Arbor: University of Michigan Press, 1994.

———. *Symbolic Exchange and Death*. Trans. Iain Hamilton Grant. London: Sage Publications, 1993.

Bauer, Candice A., and Ben Rogers. "A Billionth of a Meter Is a Big Deal." *Mechanical Engineers Today* 6, no. 1 (2004). http://www.asme.org (web pages on file with author).

Baxter, Stephen. "The Logic Pool." 1994. In *Nanotech*, ed. Jack Dann and Gardner Dozois, 153–73. New York: Ace Books, 1998.

Bear, Greg. *Blood Music*. 1985. New York: ibooks, 2002.

———. "Introduction to 'Blood Music.'" In *The Collected Stories of Greg Bear*, ed. Beth Meacham, 15–16. New York: Tor Books, 2002.

"Beauty and Brains." Directed by Joe Pytka. Copywritten by Michael Patti. Advertiser: General Electric. BBDO New York, 2003.

Benford, Gregory. "BIO/NANO/TECH." In *Nanodreams*, ed. Elton Elliott, 192–200. New York: Baen Books, 1995.

———. *Furious Gulf*. New York: Spectra, 1994.

———. *Great Sky River*. New York: Spectra, 1987.

———. *Sailing Bright Eternity*. New York: Spectra, 1995.

———. "A Scientist's Notebook: A Diamond Age." *Magazine of Fantasy and Science Fiction* 91, no. 2 (August 1996): 122–32.

———. "A Scientist's Notebook: Calculating the Future." *Magazine of Fantasy and Science Fiction* 85, no. 3 (September 1993): 78–87.

———. *Tides of Light*. New York: Spectra, 1989.

Benford, Gregory, and Elisabeth Malartre. *Beyond Human: Living with Robots and Cyborgs*. New York: Forge, 2007.

Benjamin, Walter. "The Work of Art in the Age of Mechanical Reproduction." 1935. In *Illuminations*, ed. Hannah Arendt, trans. Harry Zohn, 217–51. New York: Schocken Books, 1968.

Bensaude-Vincent, Bernadette. "Two Cultures of Nanotechnology?" *Hyle* 10 (2004): 65–82.

Berger, John. *Ways of Seeing*. London: Penguin, 1972.

Berne, Rosalyn W. *Nanotalk: Conversations with Scientists and Engineers about Ethics, Meaning, and Belief in the Development of Nanotechnology*. Mahwah, N.J.: Lawrence Erlbaum, 2006.

Berne, Rosalyn W., and Joachim Schummer. "Teaching Societal and Ethical Implications of Nanotechnology to Engineering Students through Science Fiction." *Bulletin of Science, Technology and Society* 25 (2005): 459–68.

Berube, David M. *Nano-Hype: The Truth behind the Nanotechnology Buzz*. New York: Prometheus Books, 2005.

Berube, David, and J. D. Shipman. "Denialism: Drexler vs. Roco." *IEEE Technology and Society Magazine* 23, no. 4 (2004): 22–26.

Biagioli, Mario. *Galileo's Instruments of Credit: Telescopes, Images, Secrecy*. Chicago: University of Chicago Press, 2006.

Binnig, Gerd, and Heinrich Rohrer. "In Touch with Atoms." *Reviews of Modern Physics* 71 (1999): S324–S330.

———. "Scanning Tunneling Microscopy—from Birth to Adolescence." *Reviews of Modern Physics* 59 (1987): 615–25.

Binnig, G., H. Rohrer, Ch. Gerber, and E. Weibel. "Tunneling through a Controllable Vacuum Gap." *Applied Physics Letters* 40 (1982): 178–80.

———. "7 X 7 Reconstruction on Si(111) Resolved in Real Space." *Physical Review Letters* 50 (1983): 120–23.

Blade Runner. Directed by Ridley Scott. Screenplay by Hampton Fancher and David Peoples. Warner, 1982.

Blish, James. *The Seedling Stars*. New York: Gnome Press, 1957.

———. "Surface Tension." 1952. In *The Science Fiction Hall of Fame, Volume One,*

1929–1964, ed. Robert Silverberg, 394–425. New York: Doubleday, 1970. Reprint, New York: Tor Books, 2003.

Block, Steven M. "What Is Nanotechnology?" Paper presented at the National Institutes of Health conference "Nanoscience and Nanotechnology: Shaping Biomedical Research," Natcher Conference Center, Bethesda, Md., June 25, 2000.

Bloom, Harold. *The Anxiety of Influence: A Theory of Poetry.* 1973. 2nd edition. New York: Oxford University Press, 1997.

B-M Information. "Nanotechnology in Need of Successful Communication." Zürich: Burson-Marsteller, 2004.

Boehlert, Sherwood. "Opening Statement for Hearing on Nano Consequences." U.S. House of Representatives, House Committee on Science, April 9, 2003. http://www.house.gov/science/ (visited June 25, 2005; web pages on file with author).

Bois, Yve-Alain, and Rosalind E. Krauss. *Formless: A User's Guide.* New York: Zone Books, 1997.

Bolter, Jay David, and Richard Grusin. *Remediation: Understanding New Media.* Cambridge: MIT Press, 1999.

Booker, Richard, and Earl Boysen. *Nanotechnology for Dummies.* Hoboken, N.J.: Wiley, 2005.

Borges, Jorge Luis. "On Exactitude in Science." 1946. In *Collected Fictions*, trans. Andrew Hurley, 325. New York: Penguin Books, 1998.

Borm, Paul J. A., David Robbins, Stephan Haubold, Thomas Kuhlbusch, Heinz Fissan, Ken Donaldson, Roel Schins, Vicki Stone, Wolfgang Kreyling, Jurgen Lademann, Jean Krutmann, David Warheit, and Eva Oberdörster. "The Potential Risks of Nanomaterials: A Review Carried Out for ECETOC." *Particle and Fibre Toxicology* 3 (2006): Article 11. http://www.particleandfibretoxicology.com (web pages on file with author).

Bostrom, Nick. *Anthropic Bias: Observation Selection Effects in Science and Philosophy.* New York: Routledge, 2002.

———. "Existential Risks: Analyzing Human Extinction Scenarios and Related Hazards." *Journal of Evolution and Technology* 9 (2002). http://www.jetpress.org (web pages on file with author).

Bowman, Diana M., Graeme A. Hodge, and Peter Binks. "Are We Really the Prey? Nanotechnology as Science and Science Fiction." *Bulletin of Science, Technology & Society* 22 (2007): 435–45.

Braidotti, Rosi. *Metamorphoses: Towards a Material Theory of Becoming.* Cambridge: Polity, 2002.

Brauman, John I. "Room at the Bottom." *Science* 254 (1991): 1277.

Brezinski, Darlene. "It's a Small World after All." *Paint and Coatings Industry* 14, no. 3 (2000): 12.

Brin, David. *Earth.* New York: Bantam Books, 1990.

———. "Singularities and Nightmares: The Range of Our Futures." *Nanotechnology Perceptions* 2 (2006).

Broderick, Damien. *The Spike: How Our Lives Are Being Transformed by Rapidly Advancing Technologies*. New York: Forge Books, 2001.

Brooks, Rodney Allen. *Flesh and Machines: How Robots Will Change Us*. New York: Pantheon Books, 2002.

Brown, Jim [a.k.a. The Swirling Brain]. "Vampire Bot." *Robots.net*, November 12, 2002, response to R. Steven Rainwater, "Fuel Cell Powered by Human Bodily Fluids" (November 12, 2002). http://robots.net (visited September 20, 2006; web pages on file with author).

Brown, Nathan. "Needle on the Real: Technoscience and Poetry at the Limits of Fabrication." In *Nanoculture: Implications of the New Technoscience*, ed. N. Katherine Hayles, 173–90. Bristol: Intellect Books, 2004.

Brown, Nik, Brian Rappert, and Andrew Webster, eds. *Contested Futures: A Sociology of Prospective Technoscience*. Aldershot: Ashgate, 2000.

Bruno, Lee. "The Next Big Thing Is Really Tiny: The Nanoscale World, Where Atoms Jump on Command." *Stanford* 34, no. 3 (2005): 44–49.

Bueno, Otávio. "The Drexler-Smalley Debate on Nanotechnology: Incommensurability at Work?" *Hyle* 10 (2004): 83–98.

———. "Von Neumann, Self-Reproduction, and the Constitution of Nanophenomena." In *Discovering the Nanoscale*, ed. Davis Baird, Alfred Nordmann, and Joachim Schummer, 101–15. Amsterdam: Ios Press, 2004.

Bukatman, Scott. *Terminal Identity: The Virtual Subject in Postmodern Science Fiction*. Durham: Duke University Press, 1993.

———. "There's Always Tomorrowland: Disney and the Hypercinematic Experience." *October* 57 (1991): 55–78.

Bush, Vannevar. *Science, the Endless Frontier*. Washington: United States Government Printing Office, 1945.

Butler, Judith. *Bodies That Matter: On the Discursive Limits of "Sex."* New York: Routledge, 1993.

———. *Gender Trouble: Feminism and the Subversion of Identity*. 1990. 10th anniversary edition. New York: Routledge, 1999.

Cairns-Smith, A. G. *Genetic Takeover and the Mineral Origins of Life*. Cambridge: Cambridge University Press, 1982.

Calder, Richard. *Dead Girls, Dead Boys, Dead Things*. New York: St. Martin's Griffin, 1998.

Campbell, Mary Baine. *Wonder and Science: Imagining Worlds in Early Modern Europe*. Ithaca, N.Y.: Cornell University Press, 1999.

Carter, John. *Sex and Rockets: The Occult World of Jack Parsons*. 2nd ed. Los Angeles: Feral House, 2004.

Center for Responsible Nanotechnology. "Dangers of Molecular Manufacturing." *Center for Responsible Nanotechnology*, 2003. http://www.crnano.org (visited April 15, 2005; web pages on file with author).

———. "Gray Goo Is a Small Issue." *Center for Responsible Nanotechnology*, December

14, 2003. http://www.crnano.org (visited May 5, 2005; web pages on file with author).

Clark, Andy. *Natural-Born Cyborgs: Minds, Technologies, and the Future of Human Intelligence*. Oxford: Oxford University Press, 2003.

Clarke, Arthur C. "1st Floor: Haberdashery, Curtains. 35,780th Floor: Satellite in Space." *Times* (London), September 24, 2005, Features, 25.

——. *The Fountains of Paradise*. New York: Harcourt Brace Jovanovich, 1979.

——. *Profiles of the Future: An Inquiry into the Limits of the Possible*. New York: Harper and Row, 1958.

——. "The Space Elevator: 'Thought Experiment,' or Key to the Universe?" *Advances in Earth Oriented Applications of Space Technology* 1 (1981): 39–48.

Clarke, Bruce. *Energy Forms: Allegory and Science in the Era of Classical Thermodynamics*. Ann Arbor: University of Michigan Press, 2001.

——. *Posthuman Metamorphosis: Narrative and Systems*. Bronx: Fordham University Press, 2008.

Clement, Hal. *Needle*. 1949. In *The Essential Hal Clement*, vol. 1, *Trio for Slide Rule and Typewriter*, ed. Mark L. Olson and Anthony R. Lewis, 21–201. Framingham: NESFA Press, 1999.

Clinton, William J. Address to the California Institute of Technology, January 21, 2000. Video transcript of talk produced by Caltech's Audio Visual Services, Electronic Media Publications, and Digital Media Center.

Clover, Carol. *Men, Women, and Chainsaws: Gender in the Horror Film*. Princeton: Princeton University Press, 1992.

Cobb, Michael D., and Jane Macoubrie. "Public Perceptions About Nanotechnology: Risks, Benefits, and Trust." *Journal of Nanoparticle Research* 6 (2004): 395–405.

Coenen, Christopher. "Der posthumanistische Technofuturismus in den Debatten über Nanotechnologie und Converging Technologies." In *Nanotechnologien im Kontext: Philosophische, ethische und gesellschaftliche Perspektiven*, ed. Alfred Nordmann, Joachim Schummer, and Astrid Schwarz, 195–222. Berlin: Akademische Verlagsgesellschaft, 2006.

——. "Nanofuturismus: Anmerkungen zu seiner Relevanz, Analyse und Bewertung." *Technikfolgenabschätzung—Theorie und Praxis* 13, no. 2 (2004): 78–85.

Colbert, Daniel T., and Richard E. Smalley. "Fullerene Nanotubes for Molecular Electronics." *Trends in Biotechnology* 17 (1999): 46–50.

Committee for the Review of the National Nanotechnology Initiative. *Small Wonders, Endless Frontiers: A Review of the National Nanotechnology Initiative*. Washington: National Academy Press, 2002.

Conley, Tom. *The Self-Made Map: Cartographic Writing in Early Modern France*. Minneapolis: University of Minnesota Press, 1996.

Cramer, John. "The Alternate View: Report on Nanocon 1." *Analog Science Fiction/ Science Fact* 109, no. 10 (1989): 113–17.

——. "The Alternate View: The Carbon Nanotube—Miracle Material." *Analog Science Fiction and Fact* 121, no. 12 (2001): 79–82.

————. *Einstein's Bridge*. New York: Avon Books, 1997.

————. "Nanotechnology: The Coming Storm." Foreword to *Nanodreams*, ed. Elton Elliott, 4–12. New York: Baen Books, 1995.

Crandall, B. C., ed. *Nanotechnology: Molecular Speculations on Global Abundance*. Cambridge: MIT Press, 1996.

————. Preface to *Nanotechnology: Molecular Speculations on Global Abundance*, ed. B. C. Crandall, ix–xi. Cambridge: MIT Press, 1996.

————. Preface to *Nanotechnology: Research Perspectives*, ed. B. C. Crandall and B. Lewis, vii–viii. Cambridge: MIT Press, 1997.

Crandall, B. C., and B. Lewis, eds. *Nanotechnology: Research Perspectives*. Cambridge: MIT Press, 1997.

Creed, Barbara. *The Monstrous-Feminine: Film, Feminism, Psychoanalysis*. London: Routledge, 1993.

Crichton, Michael. *Disclosure*. New York: Knopf, 1994.

————. *Prey*. New York: HarperCollins, 2002.

————. *Rising Sun*. New York: Knopf, 1992.

Crommie, M. F., C. P. Lutz, and D. M. Eigler. "Confinement of Electrons to Quantum Corrals on a Metal Surface." *Science* 262 (1993): 218–20.

Crommie, M. F., C. P. Lutz, D. M. Eigler, and E. J. Heller. "Waves on a Metal Surface and Quantum Corrals." *Surface Review and Letters* 2 (1995): 127–37.

Csányi, V., and G. Kampis. "Autogenesis: The Evolution of Replicative Systems." *Journal of Theoretical Biology* 114 (1985): 303–21.

Csicsery-Ronay, Istvan, Jr. "Futuristic Flu, or The Revenge of the Future." In *Fiction 2000: Cyberpunk and the Future of Fiction*, ed. George Slusser and Tom Shippey, 26–45. Athens: University of Georgia Press, 1992.

Cuberes, M. T., R. R. Schlittler, and J. K. Gimzewski. "Room-Temperature Repositioning of Individual C_{60} Molecules at Cu Steps: Operation of a Molecular Counting Device." *Applied Physics Letters* 69 (1996): 3016–18.

Culler, Jonathan. *The Pursuit of Signs: Semiotics, Literature, Deconstruction*. London: Routledge and Kegan Paul, 1981.

Cummings, Ray. *The Girl in the Golden Atom*. 1923. Lincoln: University of Nebraska Press, 2005.

Dai, Hongjie, Nathan Franklin, and Jie Han. "Exploiting the Properties of Carbon Nanotubes for Nanolithography." *Applied Physics Letters* 73 (1998): 1508–10.

Daston, Lorraine, and Katharine Park. *Wonders and the Order of Nature, 1150–1750*. New York: Zone Books, 1998.

Davidson, Arnold I. "The Horror of Monsters." In *The Boundaries of Humanity: Humans, Animals, Machines*, ed. James J. Sheehan and Morton Sosna, 36–67. Berkeley: University of California Press, 1991.

Davis, Anthony P. "Synthetic Molecular Motors." *Nature* 401 (1999): 120–21.

The Day the Earth Stood Still. Directed by Robert Wise. Screenplay by Edmund H. North. Twentieth Century–Fox, 1951.

Decker, Michael. "Eine Definition von Nanotechnologie: Erster Schritt für ein interdisziplinäres Nanotechnology Assessment." In *Nanotechnologien im Kontext: Philosophische, ethische und gesellschaftliche Perspektiven*, ed. Alfred Nordmann, Joachim Schummer, and Astrid Schwarz, 33–48. Berlin: Akademische Verlagsgesellschaft, 2006.

Decker, Michael, Ulrich Fiedeler, and Torsten Fleischer. "Ich sehe was, was Du nicht siehst . . . zur Definition von Nanotechnologie." *Technikfolgenabschätzung—Theorie und Praxis* 13, no. 2 (2004): 10–16.

DeGrado, W. F., L. Regan, and S. P. Ho. "The Design of a Four-Helix Bundle Protein." *Cold Spring Harbor Symposia on Quantitative Biology* 52 (1987): 521–26.

De Landa, Manuel. "Nonorganic Life." In *Incorporations*, ed. Jonathan Crary and Sanford Kwinter, 129–67. New York: Zone Books, 1992.

———. *A Thousand Years of Nonlinear History*. New York: Zone Books, 2000.

———. *War in the Age of Intelligent Machines*. New York: Zone Books, 1991.

Deleuze, Gilles. *The Fold: Leibniz and the Baroque*. 1988. Trans. Tom Conley. Minneapolis: University of Minnesota Press, 1993.

———. *Francis Bacon: The Logic of Sensation*. 1981. Trans. Daniel W. Smith. Minneapolis: University of Minnesota Press, 2003.

Deleuze, Gilles, and Félix Guattari. *Anti-Oedipus: Capitalism and Schizophrenia*. 1972. Trans. Robert Hurley, Mark Seem, and Helen R. Lane. Minneapolis: University of Minnesota Press, 1983.

———. *A Thousand Plateaus: Capitalism and Schizophrenia*. 1980. Trans. Brian Massumi. Minneapolis: University of Minnesota Press, 1987.

de Man, Paul. *Blindness and Insight: Essays in the Rhetoric of Contemporary Criticism*. 2nd rev. ed. Minneapolis: University of Minnesota Press, 1983.

Derrida, Jacques. "Deconstruction and the Other." In *Dialogues with Contemporary Continental Thinkers: The Phenomenological Heritage*, ed. Richard Kearney, 123–24. Manchester: Manchester University Press, 1984.

———. *Dissemination*. 1972. Trans. Barbara Johnson. Chicago: University of Chicago Press, 1981.

———. "The Ends of Man." 1968. In *Margins of Philosophy*, trans. Alan Bass, 109–36. Chicago: University of Chicago Press, 1982.

———. *Margins of Philosophy*. 1972. Trans. Alan Bass. Chicago: University of Chicago Press, 1982.

———. "No Apocalypse, Not Now (Full Speed Ahead, Seven Missiles, Seven Missives)." *Diacritics* 14 (1984): 20–31.

———. "Of an Apocalyptic Tone Recently Adopted in Philosophy." *Oxford Literary Review* 6, no. 2 (1984): 3–37.

———. *Of Grammatology*. 1967. Trans. Gayatri Chakravorty Spivak. Baltimore: Johns Hopkins University Press, 1976.

———. *Points . . . : Interviews, 1974–1994*. Trans. Peggy Kamuf et al. Ed. Elisabeth Weber. Stanford: Stanford University Press, 1995.

———. *The Post Card: From Socrates to Freud and Beyond*. 1980. Trans. Alan Bass. Chicago: University of Chicago Press, 1987.

————. *Specters of Marx: The State of the Debt, the Work of Mourning, and the New International*. 1993. New York: Routledge, 1994.

————. "Structure, Sign, and Play in the Discourse of the Human Sciences." 1966. In *Writing and Difference*, trans. Alan Bass, 278–93. Chicago: University of Chicago Press, 1978.

————. *Writing and Difference*. 1967. Trans. Alan Bass. Chicago: University of Chicago Press, 1978.

Dery, Mark. *Escape Velocity: Cyberculture at the End of the Century*. New York: Grove Press, 1996.

de Souza e Silva, Adriana. "The Invisible Imaginary: Museum Spaces, Hybrid Reality, and Nanotechnology." In *Nanoculture: Implications of the New Technoscience*, ed. N. Katherine Hayles, 27–46. Bristol: Intellect Books, 2004.

Dick, Philip K. "Autofac." 1955. In *Selected Stories of Philip K. Dick*, 202–26. New York: Pantheon, 2002.

Dick, Steven J. *The Biological Universe: The Twentieth-Century Extraterrestrial Life Debate and the Limits of Science*. Cambridge: Cambridge University Press, 1996.

Dinello, Daniel. *Technophobia! Science Fiction Visions of Posthuman Technology*. Austin: University of Texas Press, 2005.

Disch, Thomas M. *The Dreams Our Stuff Is Made Of: How Science Fiction Conquered the World*. New York: Simon and Schuster, 1998.

Doctorow, Cory. *Down and Out in the Magic Kingdom*. New York: Tor, 2003.

————. "The Rapture of the Geeks: Funny Hats, Transcendent Wisdom, and the Singularity." *Whole Earth Review* 111 (2003): 38–39.

Doyle, Richard. *On Beyond Living: Rhetorical Transformations of the Life Sciences*. Stanford: Stanford University Press, 1997.

————. *Wetwares: Experiments in Postvital Living*. Minneapolis: University of Minnesota Press, 2003.

Drexler, K. Eric. "Drexler Counters." *Chemical and Engineering News* 81 (2003): 40–41.

————. *Engines of Creation: The Coming Era of Nanotechnology*. 1986. 2nd ed. New York: Anchor Books, 1990.

————. "From Nanodreams to Realities." In *Nanodreams*, ed. Elton Elliott, 13–16. New York: Baen Books, 1995.

————. "Machine-Phase Nanotechnology." *Scientific American* (2001): 74–75.

————. "Molecular Engineering: An Approach to the Development of General Capabilities for Molecular Manipulation." *Proceedings of the National Academy of Sciences* 78 (1981): 5275–78.

————. "Molecular Manufacturing as a Path to Space." In *Prospects in Nanotechnology*, ed. Marcus Krummenacker and James Lewis, 197–205. New York: Wiley, 1995.

————. "Molecular Manufacturing: Perspectives on the Ultimate Limits of Fabrication." *Philosophical Transactions of the Royal Society of London A* 353 (1995): 323–31.

————. *Nanosystems: Molecular Machinery, Manufacturing, and Computation*. New York: Wiley, 1992.

————. "Nanotechnology: From Feynman to Funding." *Bulletin of Science, Technology and Society* 24 (2004): 21–27.

————. "Open Letter to Richard Smalley." *Chemical and Engineering News* 81 (2003): 38–39.

Drexler, K. Eric, Chris Peterson, and Gayle Pergamit. *Unbounding the Future: The Nanotechnology Revolution*. New York: William Morrow, 1991.

Drum, Ryan W., and Richard Gordon. "Star Trek Replicators and Diatom Nanotechnology." *Trends in Biotechnology* 21 (2003): 325–28.

Du Charme, Wesley M. *Becoming Immortal: Nanotechnology, You, and the Demise of Death*. Evergreen, Colo.: Blue Creek Ventures, 1995.

Dumit, Joseph. *Picturing Personhood: Brain Scans and Biomedical Identity*. Princeton: Princeton University Press, 2004.

Dupuy, Jean-Pierre, and Alexei Grinbaum. "Living with Uncertainty: Toward the Ongoing Normative Assessment of Nanotechnology." *Techné* 8, no. 2 (2004): 4–25.

Dyson, Freeman. "The Future Needs Us!" *The New York Review of Books* 50, no. 2 (2003): 11–13.

"Editorial: Growing Nanoknowledge." *Boston Globe*, August 27, 2002, A14.

Eigler, Donald. "From the Bottom Up: Building Things with Atoms." In *Nanotechnology*, ed. Gregory Timp, 425–35. New York: Springer, 1999.

————. "There's Plenty of Room in the Middle: A Perspective from the Bottom." The 13th Hubert M. James Lecture, Purdue University, April 4, 2001. Abstract published online by Purdue University Department of Physics, April 4, 2001. http://www.physics.purdue.edu (visited February 16, 2005; web pages on file with author).

Eigler, D. M., and E. K. Schweizer. "Positioning Single Atoms with a Scanning Tunnelling Microscope." *Nature* 344 (1990): 524–26.

Elliott, Elton. *Nanodreams*. New York: Baen Books, 1995.

————. Preface to *Nanodreams*, ed. Elton Elliott, 1–3. New York: Baen Books, 1995.

ETC Group. *The Big Down: Atomtech—Technologies Converging at the Nano-scale; From Genomes to Atoms*. Winnipeg: ETC Group, 2003.

————. "Green Goo: Nanobiotechnology Comes Alive!" *ETC Communiqué* 77 (2003): 1–10.

————. "Size Matters! The Case for a Global Moratorium." *ETC Occasional Papers Series* 7, no. 1 (2003): 1–20.

European Commission, Directorate-General for Research. *Nanotechnology: Small Science with a Big Potential*. Luxembourg: Office for Official Publications of the European Communities, 2002.

"Evolution." *Star Trek: The Next Generation*. Episode 49. Directed by Winrich Kolbe. Teleplay by Michael Piller. Paramount Television, 1989.

Faber, Brenton. "Popularizing Nanoscience: The Public Rhetoric of Nanotechnology, 1986–1999." *Technical Communication Quarterly* 15 (2006): 141–69.

Fantastic Voyage. Directed by Richard Fleischer. Screenplay by Harry Kleiner. Twentieth Century–Fox, 1966.

Farazdel, Abbas, and Michel Dupuis. "All-Electron *ab Initio* Self-Consistent-Field Study of Electron Transfer in Scanning Tunneling Microscopy at Large and Small Tip-Sample Separations: Supermolecule Approach." *Physical Review B* 44 (1991): 3909–15.

Feldstein, Al, William M. Gaines, and Harvey Kurtzman. "Lost in the Microcosm." In *Weird Science*, no. 12, 1–8. New York: EC Comics, 1950. Reprint, *The EC Archives: Weird Science*, vol. 1, ed. Russ Cochran, 11–18. Timonium, Md.: Gemstone Publishing, 2006.

Feynman, Richard. "There's Plenty of Room at the Bottom." 1959. In *Miniaturization*, ed. H. D. Gilbert, 282–96. New York: Reinhold, 1961.

———. "Tiny Machines: The Feynman Lecture on Nanotechnology." 1984. DVD. Mill Valley: Sound Photosynthesis, 2004.

Fleck, Ludwik. *Genesis and Development of a Scientific Fact*. 1935. Trans. Fred Bradley and Thaddeus J. Trenn. Ed. Thaddeus J. Trenn and Robert K. Merton. Chicago: University of Chicago Press, 1979.

Flynn, Michael. *The Nanotech Chronicles*. New York: Baen Books, 1991.

"Foresight Update 53." *Foresight Institute*, January 15, 2004. http://www.foresight.org (visited April 17, 2005; web pages on file with author).

Foster, J. S., J. E. Frommer, and P. C. Arnett. "Molecular Manipulation Using a Tunnelling Microscope." *Nature* 331 (1988): 324–26.

Foucault, Michel. *History of Madness*. 1961. Trans. Jonathan Murphy and Jean Khalfa. Ed. Jean Khalfa. London: Routledge, 2006.

———. *The Order of Things: An Archaeology of the Human Sciences*. New York: Vintage Books, 1973.

Frankel, Felice. "Capturing Quantum Corrals." *American Scientist* 93, no. 3 (2005): 261.

Franklin, H. Bruce. *War Stars: The Superweapon and the American Imagination*. New York: Oxford University Press, 1988.

Freitas, Robert A., Jr. "Molecular Manufacturing: Too Dangerous to Allow?" *Nanotechnology Perceptions* 2 (2006): 15–24.

———. *Nanomedicine*. Vol. 1, *Basic Capabilities*. Georgetown, Tex.: Landes Bioscience, 1999.

———. "Police Nanites and Nanorobot Population Limits." *sci.nanotech*, March 7, 1997. Usenet thread archived at http://groups.google.com (visited April 20, 2005; web pages on file with author).

———. "Some Limits to Global Ecophagy by Biovorous Nanoreplicators, with Public Policy Recommendations." *Foresight Institute*, April 2000. http://www .foresight.org (visited April 20, 2005; web pages on file with author).

Freitas, Robert A., and Ralph C. Merkle. *Kinematic Self-Replicating Machines*. Georgetown, Tex.: Landes Bioscience, 2004.

Freud, Sigmund. *Beyond the Pleasure Principle*. Ed. and trans. James Strachey. New York: Norton, 1961.

———. "The Uncanny." 1919. In *The Standard Edition of the Complete Psychological Works of Sigmund Freud*, vol. 17, ed. and trans. James Strachey, in collaboration with Anna Freud, assisted by Alix Strachey and Alan Tyson, 219–52. London: Hogarth Press, 1953.

Fuchs, Cynthia J. "'Death Is Irrelevant': Cyborgs, Reproduction, and the Future of Male Hysteria." In *The Cyborg Handbook*, ed. Chris Hables Gray, Heidi J. Figueroa-Sarriera, and Steven Mentor, 281–300. New York: Routledge, 1995.

Galison, Peter. *Einstein's Clocks, Poincaré's Maps: Empires of Time*. New York: W. W. Norton, 2003.

———. *Image and Logic: A Material Culture of Microphysics*. Chicago: University of Chicago Press, 1997.

Gerber, Christoph, and Hans Peter Lang. "How the Doors to the Nanoworld Were Opened." *Nature Nanotechnology* 1, no. 1 (2006): 3–5.

Gillis, Justin. "Scientists Planning to Make New Form of Life." *Washington Post*, November 21, 2002, A1.

Gimzewski, James K., and Christian Joachim. "Nanoscale Science of Single Molecules Using Local Probes." *Science* 283 (1999): 1683–88.

Gimzewski, James, and Victoria Vesna. "The Nanomeme Syndrome: The Blurring of Fact and Fiction in the Construction of a New Science." *Technoetic Arts* 1, no. 1 (2003): 7–24.

Gleick, James. *Genius: The Life and Science of Richard Feynman*. New York: Pantheon, 1992.

Glimell, Hans. "Grand Visions and Lilliput Politics: Staging the Exploration of the 'Endless Frontier.'" In *Discovering the Nanoscale*, ed. Davis Baird, Alfred Nordmann, and Joachim Schummer, 231–46. Amsterdam: IOS Press, 2004.

Goldberg, Jonathan. "Recalling Totalities: The Mirrored Stages of Arnold Schwarzenegger." In *The Cyborg Handbook*, ed. Chris Hables Gray, Heidi J. Figueroa-Sarriera, and Steven Mentor, 233–54. New York: Routledge, 1995.

Goodsell, David S. *Bionanotechnology: Lessons from Nature*. Hoboken: Wiley-Liss, 2004.

Goonan, Kathleen Ann. *Queen City Jazz*. 1994. New York: Orb, 2003.

Gorman, Michael E., James F. Groves, and Jeff Shrager. "Societal Dimensions of Nanotechnology as a Trading Zone: Results from a Pilot Project." In *Discovering the Nanoscale*, ed. Davis Baird, Alfred Nordmann, and Joachim Schummer, 63–73. Amsterdam: IOS Press, 2004.

Goscilo, Margaret. "Deconstructing *The Terminator*." *Film Criticism* 12 (1987–88): 37–52.

Graham, Elaine L., *Representations of the Post/Human: Monsters, Aliens and Others in Popular Culture*. Manchester: Manchester University Press, 2002.

Gray, Chris Hables. *Cyborg Citizen: Politics in the Posthuman Age*. New York: Routledge, 2001.

————. "'There Will Be War!': Future War Fantasies and Militaristic Science Fiction in the 1980s." *Science-Fiction Studies* 64 (1994): 315–36.

Gray, Chris Hables, Heidi J. Figueroa-Sarriera, and Steven Mentor, eds. *The Cyborg Handbook*. New York: Routledge, 1995.

Grosz, Elizabeth. *Volatile Bodies: Toward a Corporeal Feminism*. Bloomington: Indiana University Press, 1994.

Grunwald, Armin. "Nanotechnologie als Chiffre der Zukunft." In *Nanotechnologien im Kontext: Philosophische, ethische und gesellschaftliche Perspektiven*, ed. Alfred Nordmann, Joachim Schummer, and Astrid Schwarz, 49–80. Berlin: Akademische Verlagsgesellschaft, 2006.

Gubrud, Mark. "Interview with Mark Gubrud." By Sander Olson. *Nanotech.biz*, September 22, 2001. http://www.nanotech.biz (visited September 13, 2006; web pages on file with author).

Hacking, Ian. *Representing and Intervening: Introductory Topics in the Philosophy of Natural Science*. Cambridge: Cambridge University Press, 1983.

Halberstam, Judith. *Skin Shows: Gothic Horror and the Technology of Monsters*. Durham: Duke University Press, 1995.

Halberstam, Judith, and Ira Livingston, eds. *Posthuman Bodies*. Bloomington: Indiana University Press, 1995.

Hall, J. Storrs. *Nanofuture: What's Next for Nanotechnology*. Amherst, N.Y.: Prometheus Books, 2005.

————. "Overview of Nanotechnology." *sci.nanotech*, 1995. http://www.nanotech.dyndns.org (visited May 16, 2003; web pages on file with author).

————. "Utility Fog: The Stuff That Dreams Are Made Of." In *Nanotechnology: Molecular Speculations on Global Abundance*, ed. B. C. Crandall, 161–84. Cambridge: MIT Press, 1996.

————. "What I Want to Be When I Grow Up, Is a Cloud." *Extropy* 13 (1994). Reprint, *KurzweilAI.net*, July 6, 2001. http://www.kurzweilai.net (web pages on file with author).

Halperin, James L. *The First Immortal*. New York: Del Rey, 1998.

Hameroff, Stuart. *Ultimate Computing: Biomolecular Consciousness and Nanotechnology*. Amsterdam: Elsevier North-Holland, 1987.

Hamilton, Sheryl N. "Traces of the Future: Biotechnology, Science Fiction, and the Media." *Science Fiction Studies* 30 (2003): 267–79.

Hanafi, Zakiya. *The Monster in the Machine: Magic, Medicine, and the Marvelous in the Time of the Scientific Revolution*. Durham: Duke University Press, 2000.

Hansen, Mark. *Embodying Technesis: Technology beyond Writing*. Ann Arbor: University of Michigan Press, 2000.

————. *New Philosophy for New Media*. Cambridge: MIT Press, 2004.

Hanson, Robin. "A Critical Discussion of Vinge's Singularity Concept: Thirteen Comments." *Extropy: Journal of Transhumanist Solutions*, October 1998. http://spock.extropy.org (visited September 13, 2006; web pages on file with author).

Hanson, Valerie Louise. "Haptic Visions: Rhetorics, Subjectivities and Visualization

Technologies in the Case of the Scanning Tunneling Microscope." Ph.D. diss., Pennsylvania State University, 2004.

Hansson, Sven Ove. "Great Uncertainty about Small Things." *Techné* 8, no. 2 (2004): 26–35.

Haraway, Donna J. "A Cyborg Manifesto: Science, Technology, and Socialist-Feminism in the Late Twentieth Century." 1985. In *Simians, Cyborgs, and Women: The Reinvention of Nature*, 149–81. New York: Routledge, 1991.

———. *Modest_Witness@ Second_Millennium.FemaleMan©_Meets_OncoMouse™: Feminism and Technoscience*. New York: Routledge, 1997.

———. *Primate Visions: Gender, Race, and Nature in the World of Modern Science*. New York: Routledge, 1989.

———. "The Promises of Monsters: A Regenerative Politics for Inappropriate/d Others." In *Cultural Studies*, ed. Lawrence Grossberg, Cary Nelson, and Paula A. Treichler, 295–337. New York: Routledge, 1992.

———. *Simians, Cyborgs, and Women: The Reinvention of Nature*. New York: Routledge, 1991.

———. "Situated Knowledges: The Science Question in Feminism and the Privilege of Partial Perspective." 1988. In *Simians, Cyborgs, and Women: The Reinvention of Nature*, 183–201. New York: Routledge, 1991.

Hardt, Michael, and Antonio Negri. *Empire*. Cambridge: Harvard University Press, 2000.

Harrow, Jeffrey. "A Hundred Billion Billion Bytes—in a Sugar Cube?" *Hardware Central*, May 31, 1999. http://www.hardwarecentral.com (visited February 12, 2005; web pages on file with author).

———. "Of Numbers and Nature: Size Matters." *Harrow Technology Report*, September 15, 2003. http://www.theharrowgroup.com (visited April 17, 2005; web pages on file with author).

Hawking, Stephen. *A Brief History of Time: From the Big Bang to Black Holes*. New York: Bantam Books, 1988.

Hayles, N. Katherine. "Flesh and Metal: Reconfiguring the Mindbody in Virtual Environments." *Configurations* 10 (2002): 297–320.

———. "Gender Encoding in Fluid Mechanics: Masculine Channels and Feminine Flows." *Differences* 4, no. 2 (1992): 16–44.

———. *How We Became Posthuman: Virtual Bodies in Cybernetics, Literature, and Informatics*. Chicago: University of Chicago Press, 1999.

———. *My Mother Was a Computer: Digital Subjects and Literary Texts*. Chicago: University of Chicago Press, 2005.

———, ed. *Nanoculture: Implications of the New Technoscience*. Bristol: Intellect Books, 2004.

Heckl, Wolfgang M. "Molecular Self-Assembly and Nanomanipulation—Two Key Technologies in Nanoscience and Templating." *Advanced Engineering Materials* 6 (2004): 843–47.

Heidegger, Martin. "The Age of the World Picture." 1938. In *The Question Concerning*

Technology and Other Essays, trans. William Lovitt, 115–54. New York: Garland, 1977.

———. "The Question Concerning Technology." 1955. In *The Question Concerning Technology and Other Essays*, trans. William Lovitt, 3–35. New York: Garland, 1977.

———. *The Question Concerning Technology and Other Essays*. Trans. William Lovitt. New York: Garland, 1977.

Heinlein, Robert A. *Stranger in a Strange Land*. New York: Putnam, 1961.

———. "Waldo." 1942. In *Waldo and Magic, Inc.*, by Robert A. Heinlein, 1–154. New York: Del Rey, 1986.

———. *Waldo and Magic, Inc.* 1950. New York: Del Rey, 1986.

Heinrich, A. J., C. P. Lutz, J. A. Gupta, and D. M. Eigler. "Molecule Cascades." *Science* 298 (2002): 1381–87.

Hellraiser. Written and directed by Clive Barker. New World Pictures, 1987.

Helmreich, Stefan. *Silicon Second Nature: Culturing Artificial Life in a Digital World*. Updated ed. Berkeley: University of California Press, 1998.

Henderson, Mark. "Prince's 'Goo' Fears Are Nonsense, Say Experts." *Times* (London), April 29, 2003, Home News, 11.

Hennig, Jochen. "Changes in the Design of Scanning Tunneling Microscope Images from 1980 to 1990." *Techné* 8, no. 2 (2004): 36–55.

Hessenbruch, Arne. "Beyond Truth: Pleasure of Nanofutures." *Techné* 8, no. 3 (2005): 34–61.

———. "Nanotechnology and the Negotiation of Novelty." In *Discovering the Nanoscale*, ed. Davis Baird, Alfred Nordmann, and Joachim Schummer, 135–44. Amsterdam: IOS Press, 2004.

Hissink, Dennis. "Nanotechnology Makes a Small World Even Smaller." *LetsGoDigital*, October 29, 2004. http://www.letsgodigital.net (visited February 12, 2005; web pages on file with author).

Ho, Dean, Andrew O. Fung, and Carlo D. Montemagno. "Engineering Novel Diagnostic Modalities and Implantable Cytomimetic Nanomaterials for Next-Generation Medicine." *Biology of Blood and Marrow Transplantation* 12 (2006): 92–99.

HRH the Prince of Wales [Charles Windsor]. "Menace in the Minutiae: New Nanotechnology Has Potential Dangers as Well as Benefits." *Independent on Sunday*, July 11, 2004, 3, 25.

Huet, Marie-Hélène. *Monstrous Imagination*. Cambridge: Harvard University Press, 1993.

Hurley, Kelly. "Reading like an Alien: Posthuman Identity in Ridley Scott's *Alien* and David Cronenberg's *Rabid*." In *Posthuman Bodies*, ed. Judith Halberstam and Ira Livingston, 203–24. Bloomington: Indiana University Press, 1995.

IBM Almaden Visualization Lab. "The Corral Reef." *STM Image Gallery, IBM Almaden Research Center Visualization Lab*, 1995. http://www.almaden.ibm.com/ (visited March 15, 2005; web pages on file with author).

Ihde, Don. *Bodies in Technology*. Minneapolis: University of Minnesota Press, 2002.

————. *Technology and the Lifeworld: From Garden to Earth.* Bloomington: Indiana University Press, 1990.

Irigaray, Luce. *Speculum of the Other Woman.* 1974. Trans. Gillian C. Gill. Ithaca, N.Y.: Cornell University Press, 1985.

————. *This Sex Which Is Not One.* 1977. Trans. Catherine Porter. Ithaca, N.Y.: Cornell University Press, 1985.

Jacobstein, Neil. "Foresight Guidelines for Responsible Nanotechnology Development" (Version 6.0). *Foresight Institute,* 2005. http://www.foresight.org (visited March 15, 2005; web pages on file with author).

Jameson, Fredric. *Postmodernism, or, the Cultural Logic of Late Capitalism.* Durham: Duke University Press, 1991.

Jason X. Directed by James Isaac. Screenplay by Todd Farmer. New Line, 2001.

Jay, Martin. *Downcast Eyes: The Denigration of Vision in Twentieth-Century French Thought.* Berkeley: University of California Press, 1993.

Johnson, Ann. "The End of Pure Science: Science Policy from Bayh-Dole to the NNI." In *Discovering the Nanoscale,* ed. Davis Baird, Alfred Nordmann, and Joachim Schummer, 217–29. Amsterdam: IOS Press, 2004.

————. "The Shape of Molecules to Come." In *Simulation: Pragmatic Construction of Reality,* ed. Johannes Lenhard, Günter Küppers, and Terry Shinn, 25–39. Dordrecht: Springer, 2006.

Johnson, Barbara. "The Frame of Reference: Poe, Lacan, Derrida." In "Literature and Psychoanalysis: The Question of Reading—Otherwise," special issue, *Yale French Studies* 55–56 (1977): 457–505.

————. *A World of Difference.* 1987. Baltimore: Johns Hopkins University Press, 1989.

Johnston, John. "A Future for Autonomous Agents: Machinic *Merkwelten* and Artificial Evolution." *Configurations* 10 (2002): 473–516.

————. *Information Multiplicity: American Fiction in the Age of Media Saturation.* Baltimore: Johns Hopkins University Press, 1998.

————. "Machinic Vision." *Critical Inquiry* 26 (1999): 27–48.

Johnston, Sean. *Holographic Visions: A History of New Science.* Oxford: Oxford University Press, 2006.

Jones, David E. H. "Technical Boundless Optimism." *Nature* 374 (1995): 835–37.

Jones, Richard A. L. *Soft Machines: Nanotechnology and Life.* Oxford: Oxford University Press, 2004.

Jordanova, Ludmilla. *Sexual Visions: Images of Gender in Science and Medicine between the Eighteenth and Twentieth Centuries.* Madison: University of Wisconsin Press, 1989.

Joy, Bill. "Why the Future Doesn't Need Us." *Wired* 8, no. 4 (2000): 238–63.

Junk, Andreas, and Falk Riess. "From an Idea to a Vision: There's Plenty of Room at the Bottom." *American Journal of Physics* 74 (2006): 825–30.

Kaiser, Mario. "Drawing the Boundaries of Nanoscience—Rationalizing the Concerns?" *Journal of Law, Medicine and Ethics* 34 (2006): 667–74.

Kakoudaki, Despina. "Pinup and Cyborg: Exaggerated Gender and Artificial Intelligence." In *Future Females: The Next Generation*, ed. Marleen S. Barr, 165–95. Lanham, Md.: Rowman and Littlefield, 2000.

Kauffman, Stuart A. *At Home in the Universe: The Search for Laws of Self-Organization and Complexity*. Oxford: Oxford University Press, 1995.

———. *Investigations*. Oxford: Oxford University Press, 2000.

Kay, Lily E. *The Molecular Vision of Life: Caltech, the Rockefeller Foundation, and the Rise of the New Biology*. Oxford: Oxford University Press, 1993.

———. *Who Wrote the Book of Life? A History of the Genetic Code*. Stanford: Stanford University Press, 2000.

Kearnes, Matthew. "Chaos and Control: Nanotechnology and the Politics of Emergence." *Paragraph* 29, no. 2 (2006): 57–80.

Keller, Evelyn Fox. *The Century of the Gene*. Cambridge: Harvard University Press, 2000.

———. *Reflections on Gender and Science*. 1985. 10th anniversary edition. New Haven: Yale University Press, 1995.

Kelly, Kevin. *Out of Control: The New Biology of Machines, Social Systems and the Economic World*. Cambridge: Perseus Books, 1994.

Kern, Louis J. "American 'Grand Guignol': Splatterpunk Gore, Sadean Morality and Socially Redemptive Violence." *Journal of American Culture* 19 (1996): 47–59.

Khushf, George. "A Hierarchical Architecture for Nano-scale Science and Technology: Taking Stock of the Claims about Science Made by Advocates of NBIC Convergence." In *Discovering the Nanoscale*, ed. Davis Baird, Alfred Nordmann, and Joachim Schummer, 21–33. Amsterdam: IOS Press, 2004.

Kilgore, De Witt Douglas. *Astrofuturism: Science, Race, and Visions of Utopia in Space*. Philadelphia: University of Pennsylvania Press, 2003.

Kim, Philip, and Charles M. Lieber. "Nanotube Nanotweezers." *Science* 286 (1999): 2148–50.

Kittler, Friedrich. *Gramophone, Film, Typewriter*. Stanford: Stanford University Press, 1999.

Knight, Damon Francis. *In Search of Wonder: Essays on Modern Science Fiction*. 1956. 3rd ed. Chicago: Advent Publishers, 1996.

Knight, Jonathan. "The Engine of Creation." *New Scientist* 162, no. 2191 (1999): 38–41.

Köchy, Kristian. "Maßgeschneiderte nanoskalige Systeme: Methodologische und ontologische Überlegungen." In *Nanotechnologien im Kontext: Philosophische, ethische und gesellschaftliche Perspektiven*, ed. Alfred Nordmann, Joachim Schummer, and Astrid Schwarz, 131–50. Berlin: Akademische Verlagsgesellschaft, 2006.

Koontz, Dean. *Midnight*. New York: Putnam, 1989.

Kornberg, Arthur. *For the Love of Enzymes: The Odyssey of a Biochemist*. Cambridge: Harvard University Press, 1989.

Kosko, Bart. "Despite Skeptics and Critics, Cryonics May Be a Cool Way to Go." *Los Angeles Times*, July 19, 2002, B15.

————. *The Fuzzy Future: From Society and Science to Heaven in a Chip.* New York: Harmony Books, 1999.

————. *Fuzzy Thinking: The New Science of Fuzzy Logic.* New York: Hyperion, 1993.

————. *Nanotime.* New York: Avon Books, 1997.

Kristeva, Julia. *Powers of Horror: An Essay on Abjection.* Trans. Leon S. Roudiez. New York: Columbia University Press, 1982.

Krummenacker, Markus. "Steps towards Molecular Manufacturing." *Chemical Design Automation News* 9 (1994): 1, 29–39.

Krummenacker, Markus, and James Lewis, eds. *Prospects in Nanotechnology: Toward Molecular Manufacturing.* New York: Wiley, 1995.

Kuhn, Annette, ed. *Alien Zone: Cultural Theory and Contemporary Science Fiction Cinema.* London: Verso, 1990.

Kuhn, Thomas S. *The Structure of Scientific Revolutions.* 1962. 3rd edition. Chicago: University of Chicago Press, 1996.

Kulinowski, Kristen. "Nanotechnology: From 'Wow' to 'Yuck'?" *Bulletin of Science, Technology and Society* 24 (2004): 13–20.

Kurath, Monika, and Sabine Maasen. "Toxicology as a Nanoscience?—Disciplinary Identities Reconsidered." *Particle and Fibre Toxicology* 3 (2006): Article 6. http://www.particleandfibretoxicology.com (web pages on file with author).

Kurzweil, Ray. "After the Singularity: A Talk with Ray Kurzweil." 2002. In *The Ray Kurzweil Reader: A Collection of Essays by Ray Kurzweil Published on KurzweilAI. Net, 2001–2003,* ed. Ray Kurzweil, 141–51. KurzweilAI, 2003.

————. *The Age of Intelligent Machines.* Cambridge: MIT Press, 1990.

————. *The Age of Spiritual Machines: When Computers Exceed Human Intelligence.* New York: Viking, 1999.

————, ed. *The Ray Kurzweil Reader: A Collection of Essays by Ray Kurzweil Published on KurzweilAI.Net, 2001–2003.* KurzweilAI, 2003.

————. *The Singularity Is Near: When Humans Transcend Biology.* New York: Viking, 2005.

Kurzweil, Ray, and Terry Grossman. *Fantastic Voyage: Live Long Enough to Live Forever.* Emmaus, Pa.: Rodale Press, 2004.

Kurzweil, Ray, and Max More. "Max More and Ray Kurzweil on the Singularity." 2002. In *The Ray Kurzweil Reader: A Collection of Essays by Ray Kurzweil Published on KurzweilAI.Net, 2001–2003,* ed. Ray Kurzweil, 154–70. KurzweilAI.net, 2003.

Lacan, Jacques. *Écrits.* 1966. Trans. Bruce Fink, in collaboration with Héloise Fink and Russell Grigg. New York: Norton, 2006.

————. "The Mirror Stage as Formative of the *I* Function as Revealed in Psychoanalytic Experience." 1949. In *Écrits,* trans. Bruce Fink, in collaboration with Héloise Fink and Russell Grigg, 75–81. New York: Norton, 2006.

————. *The Seminar of Jacques Lacan, Book I: Freud's Papers on Technique, 1953–1954.* 1975. Trans. John Forrester. Ed. Jacques-Alain Miller. New York: Norton, 1991.

————. *The Seminar of Jacques Lacan, Book III: The Psychoses, 1955–1956.* 1981. Trans. Russell Grigg. Ed. Jacques-Alain Miller. London: Routledge, 1993.

—————. *The Seminar of Jacques Lacan, Book XI: The Four Fundamental Concepts of Psychoanalysis.* Translated by Alan Sheridan. Edited by Jacques-Alain Miller. New York: Norton, 1998.

—————. *The Seminar of Jacques Lacan, Book XX, Encore, 1972–1973: On Feminine Sexuality; The Limits of Love and Knowledge.* 1975. Trans. Bruce Fink. Ed. Jacques-Alain Miller. New York: Norton, 1998.

—————. "Seminar on 'The Purloined Letter.'" 1955. In *Écrits*, trans. Bruce Fink, in collaboration with Héloise Fink and Russell Grigg, 6–48. New York: Norton, 2006.

—————. "The Signification of the Phallus." 1958. In *Écrits*, trans. Bruce Fink, in collaboration with Héloise Fink and Russell Grigg, 575–84. New York: Norton, 2006.

Landon, Brooks. "Less Is More: Much Less Is Much More: The Insistent Allure of Nanotechnology Narratives in Science Fiction Literature." In *Nanoculture: Implications of the New Technoscience*, ed. N. Katherine Hayles, 132–46. Bristol: Intellect Books, 2004.

—————. *Science Fiction after 1900: From the Steam Man to the Stars.* New York: Routledge, 1995.

Langton, Chris. *Artificial Life.* Redwood City, Calif.: Addison-Wesley, 1989.

Laqueur, Thomas. *Making Sex: Body and Gender from the Greeks to Freud.* Cambridge: Harvard University Press, 1990.

Latour, Bruno. *We Have Never Been Modern.* Trans. Catherine Porter. Cambridge: Harvard University Press, 1993.

Laurent, Louis, and Jean-Claude Petit. "Nanosciences and Its Convergence with Other Technologies: New Golden Age or Apocalypse?" *Hyle* 11 (2005): 45–76.

Lee, H. J., and W. Ho. "Single-Bond Formation and Characterization with a Scanning Tunneling Microscope." *Science* 286 (1999): 1719–22.

Lee, Ian Y., Xiaolei Liu, Bart Kosko, and Chongwu Zhou. "Nanosignal Processing: Stochastic Resonance in Carbon Nanotubes That Detect Subthreshold Signals." *Nano Letters* 3 (2003): 1683–86.

Lehn, Jean-Marie. "Supramolecular Chemistry: From Molecular Information towards Self-Organization and Complex Matter." *Reports on Progress in Physics* 67 (2004): 249–65.

—————. "Toward Complex Matter: Supramolecular Chemistry and Self-Organization." *PNAS* 99, no. 8 (2002): 4763–68.

Lenhard, Johannes. "Nanoscience and the Janus-Faced Character of Simulations." In *Discovering the Nanoscale*, ed. Davis Baird, Alfred Nordmann, and Joachim Schummer, 93–100. Amsterdam: IOS Press, 2004.

Lenhard, Johannes, Günter Küppers, and Terry Shinn, eds. *Simulation: Pragmatic Construction of Reality.* Dordrecht: Springer, 2006.

Lenoir, Timothy. *Instituting Science: The Cultural Production of Scientific Disciplines.* Stanford: Stanford University Press, 1997.

Levinas, Emmanuel. *Totality and Infinity: An Essay on Exteriority.* 1961. Trans. Alphonso Lingis. Pittsburgh: Duquesne University Press, 1969.

Lévi-Strauss, Claude. *The Elementary Structures of Kinship.* 1949. Rev. ed. Trans. James H. Bell, John R. von Sturmer, and Rodney Needham. Boston: Beacon Press, 1969.

Lewak, Susan E. "What's the Buzz? Tell Me What's a-Happening: Wonder, Nanotechnology, and Alice's Adventures in Wonderland." In *Nanoculture: Implications of the New Technoscience,* ed. N. Katherine Hayles, 201–10. Bristol: Intellect Books, 2004.

Lewenstein, Bruce V. "What Counts as a 'Social and Ethical Issue' in Nanotechnology?" *Hyle* 11 (2005): 5–18.

López, José. "Bridging the Gaps: Science Fiction in Nanotechnology." *Hyle* 10 (2004): 129–52.

———. "Compiling the Ethical, Legal and Social Implications of Nanotechnology." *Health Law Review* 12, no. 3 (2004): 24–27.

Lösch, Andreas. "Anticipating the Futures of Nanotechnology: Visionary Images as Means of Communication." *Technology Analysis and Strategic Management* 18 (2006): 393–409.

———. "Means of Communicating Innovations: A Case Study for the Analysis and Assessment of Nanotechnology's Futuristic Visions." *Science, Technology & Innovation Studies* 2 (2006): 103–25.

———. "Nanomedicine and Space: Discursive Orders of Mediating Innovations." In *Discovering the Nanoscale,* ed. Davis Baird, Alfred Nordmann, and Joachim Schummer, 193–202. Amsterdam: Ios Press, 2004.

Love Story. Directed by Arthur Hiller. Screenplay by Erich Segal. Paramount, 1970.

Lovejoy, Alan. "Re: Miscellaneous." *sci.nanotech,* May 31, 1989. Usenet thread archived at http://groups.google.com/group/sci.nanotech/ (visited April 20, 2005; web pages on file with author).

Lovy, Howard. "Drexler on 'Drexlerians.'" *Howard Lovy's Nanobot,* December 15, 2003. http://nanobot.blogspot.com (visited April 17, 2005, web pages on file with author).

———. "The Great Little Balls of Britain." *Howard Lovy's Nanobot,* September 7, 2007. http://nanobot.blogspot.com (visited December 8, 2007; web pages on file with author).

———. "National Nanotechnology 'Disarmament' Initiative." *Howard Lovy's Nanobot,* March 9, 2004. http://nanobot.blogspot.com (visited April 17, 2005; web pages on file with author).

———. "Welcome to Nano Reality TV, Where the Show Is Mistaken for Truth." *Small Times,* August 29, 2003. http://www.smalltimes.com (visited April 26, 2005; web pages on file with author).

Lyotard, Jean-François. *The Postmodern Condition: A Report on Knowledge.* Trans. Geoff Bennington and Brian Massumi. Minneapolis: University of Minnesota Press, 1984.

MacKinnon, Kenneth. *Love, Tears, and the Male Spectator.* Madison, N.J.: Fairleigh Dickinson University Press, 2002.

Mamin, H. J., S. Chiang, H. Birk, P. H. Guether, and D. Rugar. "Gold Deposition from a Scanning Tunneling Microscope Tip." *Journal of Vacuum Science and Technology B* 9 (1991): 1398–1402.

Maniloff, Jack. "Nannobacteria: Size Limits and Evidence." *Science* 276 (1997): 1776.

Mano, Nicholas, Fei Mao, and Adam Heller. "A Miniature Biofuel Cell Operating in a Physiological Buffer." *Journal of the American Chemical Society* 124 (2002): 12962–63.

Manovich, Lev. "The Anti-sublime Ideal in Data Art." 2002. http://www.manovich .net. Originally published as "The Anti-sublime Ideal in New Media," in the online journal *Chair et métal* 7 (2002).

———. *The Language of New Media.* Cambridge: MIT Press, 2002.

Markley, Robert. *Dying Planet: Mars in Science and the Imagination.* Durham: Duke University Press, 2005.

Marlow, John Robert. *Nano.* New York: Forge, 2004.

———. "The Sound of Inevitability—Why Nanotech *Will* Happen." *Nanoveau* 002 (2004). http://www.nanoveau.com (visited March 16, 2005; web pages on file with author).

Marshall, Kate. "Future Present: Nanotechnology and the Scene of Risk." In *Nanoculture: Implications of the New Technoscience,* ed. N. Katherine Hayles, 147–59. Bristol: Intellect Books, 2004.

Martin, Emily. "The Egg and the Sperm: How Science Has Constructed a Romance Based on Stereotypical Male-Female Roles." *Signs* 16 (1991): 485–501.

Marton, Ladislaus. "Alice in Electronland." *American Scientist* 31 (1943): 247–54.

Mason, Carol. "Terminating Bodies: Toward a Cyborg History of Abortion." In *Posthuman Bodies,* ed. Judith Halberstam and Ira Livingston, 225–43. Bloomington: Indiana University Press, 1995.

Massumi, Brian. *Parables for the Virtual: Movement, Affect, Sensation.* Durham: Duke University Press, 2002.

Matheson, Richard. *The Shrinking Man.* 1956. Reprinted as *The Incredible Shrinking Man.* New York: Tor, 1995.

The Matrix. Directed by the Wachowski Brothers. Screenplay by the Wachowski Brothers. Warner, 1999.

McAuley, Paul J. *Fairyland.* New York: Avon Books, 1995.

McCarthy, Wil. *Bloom.* New York: Del Rey, 1998. Paperback reprint, 1999.

———. *The Collapsium.* New York: Del Rey, 2000.

———. *Hacking Matter: Levitating Chairs, Quantum Mirages, and the Infinite Weirdness of Programmable Atoms.* New York: Basic Books, 2003.

———. *Murder in the Solid State.* New York: Tor, 1996.

———. "Programmable Matter: How Quantum Wellstone Ushered in Our Modern Age." *Nature* 407 (2000): 127.

———. *The Wellstone*. New York: Bantam Books, 2003.

McCarty, John. *Splatter Movies: Breaking the Last Taboo of the Screen*. New York: St. Martin's Press, 1984.

McCray, W. Patrick. "Will Small Be Beautiful? Making Policies for Our Nanotech Future." *History and Technology* 21 (2005): 177–203.

McDonald, Ian. *Evolution's Shore*. Trade paperback edition. New York: Bantam Books, 1995.

———. *Necroville*. London: Gollancz, 1994.

McHale, Brian. *Postmodernist Fiction*. New York: Routledge, 1997.

McKendree, Tom. "Nanotech Hobbies." In *Nanotechnology*, ed. B. C. Crandall, 135–44. Cambridge: MIT Press, 1996.

McKibben, Bill. *Enough: Staying Human in an Engineered Age*. New York: Times Books, 2003.

McLuhan, Marshall. *Understanding Media: The Extensions of Man*. 1964. Cambridge: MIT Press, 1994.

McRoy, Jay. "There Are No Limits: Splatterpunk, Clive Barker, and the Body *in extremis*." *Paradoxa* 17 (2002): 130–50.

Mehta, Michael D. "The Future of Nanomedicine Looks Promising, but Only If We Learn from the Past." *Health Law Review* 13, no. 1 (2004): 16–18.

———. "Some Thoughts on the Economic Impacts of Assembler-Era Nanotechnology." *Health Law Review* 12, no. 3 (2004): 33–36.

Mellor, Felicity. "Between Fact and Fiction: Demarcating Science from Non-science in Popular Physics Books." *Social Studies of Science* 33 (2003): 509–38.

Merchant, Carolyn. *The Death of Nature: Women, Ecology, and the Scientific Revolution*. New York: Harper and Row, 1980.

Merkle, Ralph C. "Cryonics." Merkle.com, March 23, 2005. http://www.merkle.com/cryo (visited July 30, 2006; web pages on file with author).

———. "Design-Ahead for Nanotechnology." In *Prospects in Nanotechnology*, ed. Marcus Krummenacker and James Lewis, 23–52. New York: Wiley, 1995.

———. "It's a Small, Small, Small, Small World." *Technology Review* 100, no. 2 (1997): 25–32.

———. "Letter to the Editor." *Technology Review* 102, no. 3 (1999): 15–16.

———. "Molecular Manufacturing: Adding Positional Control to Chemical Synthesis." *Chemical Design Automation News* 8 (1993): 1, 55–61.

———. "Nanotechnology and Medicine." In *Advances in Anti-aging Medicine*, ed. Ronald M. Klatz and Francis A. Kovarik, 277–86. Larchmont, N.Y.: Liebert, 1996.

———. "A Response to *Scientific American*'s News Story 'Trends in Nanotechnology.'" *Foresight Institute*, March 19, 1996. http://www.islandone.org (visited May 13, 2005; web pages on file with author).

———. "Self-Replicating Systems and Molecular Manufacturing." *Journal of the British Interplanetary Society* 45 (1992): 407–13.

Metz, Christian. *The Imaginary Signifier: Psychoanalysis and the Cinema.* Bloomington: Indiana University Press, 1982.

Meyer, Martin, and Osmo Kuusi. "Nanotechnology: Generalizations in an Interdisciplinary Field of Science and Technology." *Hyle* 10 (2004): 153–68.

Miksanek, Tony. "Microscopic Doctors and Molecular Black Bags: Science Fiction's Prescription for Nanotechnology and Medicine." *Literature and Medicine* 20 (2001): 55–70.

Milburn, Colin. "Monsters in Eden: Darwin and Derrida." *MLN* 118 (2003): 603–21.

———. "Nanotechnology in the Age of Posthuman Engineering: Science Fiction as Science." *Configurations* 10 (2002): 261–95.

———. "Nanowarriors: Military Nanotechnology and Comic Books." *Intertexts* 9 (2005): 77–103.

Milburn, Gerard J. *The Feynman Processor: Quantum Entanglement and the Computing Revolution.* Reading, Mass.: Perseus Books, 1998.

———. *Schrödinger's Machines: The Quantum Technology Reshaping Everyday Life.* New York: W. H. Freeman, 1997.

Miller, Virginia M., George Rodgers, Jon A. Charlesworth, Brenda Kirkland, Sandra R. Severson, Todd E. Rasmussen, Marineh Yagubyan, Jeri C. Rodgers, Franklin R. Cockerill, Robert L. Folk, Vivek Kumar, Gerard Farell-Baril, and John C. Lieske. "Evidence of Nanobacterial-Like Structures in Human Calcified Arteries and Cardiac Valves." *American Journal of Physiology—Heart and Circulatory Physiology* 287 (2004): H1115–H1124.

Mirkin, Chad A. "Tweezers for the Nanotool Kit." *Science* 286 (1999): 2095–96.

Missomelius, Petra. "Visualisierungstechniken: Die medial vermittelte Sicht auf die Welt in Kunst und Wissenschaft." In *Nanotechnologien im Kontext: Philosophische, ethische und gesellschaftliche Perspektiven,* ed. Alfred Nordmann, Joachim Schummer, and Astrid Schwarz, 169–78. Berlin: Akademische Verlagsgesellschaft, 2006.

Mnyusiwalla, Anisa, Abdallah S. Daar, and Peter A. Singer. "'Mind the Gap': Science and Ethics in Nanotechnology." *Nanotechnology* 14 (2003): R9–R13.

Mody, Cyrus C. M. "How Probe Microscopists Became Nanotechnologists." In *Discovering the Nanoscale,* ed. Davis Baird, Alfred Nordmann, and Joachim Schummer, 119–33. Amsterdam: ios Press, 2004.

———. "Small, but Determined: Technological Determinism in Nanoscience." *Hyle* 10 (2004): 99–128.

Modzelewski, Mark. "Industry Can Help Groundbreaking Nanotech Bill Fulfill Its Promise." *Small Times,* January 26, 2004. http://www.smalltimes.com (visited October 20, 2006; web pages on file with author).

Montemagno, Carlo D. "Nanomachines: A Roadmap for Realizing the Vision." *Journal of Nanoparticle Research* 3 (2001): 1–3.

Moor, James, and John Weckert. "Nanoethics: Assessing the Nanoscale from an Ethical Point of View." In *Discovering the Nanoscale,* ed. Davis Baird, Alfred Nordmann, and Joachim Schummer, 301–10. Amsterdam: ios Press, 2004.

Moravec, Hans. *Mind Children: The Future of Robot and Human Intelligence.* Cambridge: Harvard University Press, 1988.

———. Review of *Engines of Creation*, by K. Eric Drexler. *Technology Review* 89, no. 7 (1986): 76–77.

———. *Robot: Mere Machine to Transcendent Mind.* Oxford: Oxford University Press, 1999.

———. "Singularity Equation Correction." 1999. In *The Ray Kurzweil Reader: A Collection of Essays by Ray Kurzweil Published on KurzweilAI.Net, 2001–2003*, ed. Ray Kurzweil, 137–40. KurzweilAI, 2003.

More, Max. "The Extropian Principles 3.0: A Transhumanist Declaration." *MaxMore .com*, 1998. http://www.maxmore.com (visited September 14, 2005; web pages on file with author).

Morton, Timothy. "Wordsworth Digs the Lawn." *European Romantic Review* 15 (2004): 317–27.

Mullen, R. D. "Two Early Works by Ray Cummings: 'The Fire People' and 'Around the Universe.'" *Science Fiction Studies* 26 (1999): 295–302.

Muller, John P., and William J. Richardson, eds. *The Purloined Poe: Lacan, Derrida, and Psychoanalytic Reading.* Baltimore: Johns Hopkins University Press, 1988.

Mulvey, Laura. *Visual and Other Pleasures.* Bloomington: Indiana University Press, 1989.

———. "Visual Pleasure and Narrative Cinema." 1973. In *Visual and Other Pleasures*, 14–26. Bloomington: Indiana University Press, 1989.

Myers, Greg. "Scientific Speculation and Literary Style in a Molecular Genetics Article." *Science in Context* 4 (1991): 321–46.

Nagata, Linda. *Limit of Vision.* New York: Tor Books, 2001.

Nano Breaker. Directed by Kenichiro Kato. Sony PlayStation 2 video game. Konami, 2005.

"The nanoManipulator: A Virtual-Reality Interface to Scanned-Probe Microscopes." University of North Carolina, Chapel Hill: Computer Science, Physics and Astronomy, and Applied and Materials Science, 2002. http://www.cs.unc.edu (pdf pages on file with author).

"Nanotechnology's Unhappy Father." *The Economist (The Economist Technology Quarterly)* 370, no. 8366 (2004): 41–42.

National Nanotechnology Initiative. "Frequently Asked Questions" (2003 update). *National Nanotechnology Initiative*, 2003. http://www.nano.gov (visited April 25, 2004; web pages on file with author).

National Research Council, Committee to Review the National Nanotechnology Initiative. *A Matter of Size: Triennial Review of the National Nanotechnology Initiative.* Washington: National Academies Press, 2006.

National Science and Technology Council. *Nanotechnology: Shaping the World Atom by Atom.* Washington: National Science and Technology Council, 1999.

———. *National Nanotechnology Initiative: Leading to the Next Industrial Revolution.* Washington: Office of Science and Technology Policy, 2000.

Nerlich, Brigitte. "From Nautilus to Nanobo(a)ts: The Visual Construction of Nanoscience." *AZojono: Journal of Nanotechnology Online* (2005): 10.2240/azojono0109. http://www.azonano.com (web pages on file with author).

Nordmann, Alfred. "Molecular Disjunctions: Staking Claims at the Nanoscale." In *Discovering the Nanoscale*, ed. Davis Baird, Alfred Nordmann, and Joachim Schummer, 51–62. Amsterdam: IOS Press, 2004.

———. "Nanotechnology's Worldview: New Space for Old Cosmologies." *IEEE Technology and Society Magazine* 23, no. 4 (2004): 48–54.

———. "*Noumenal* Technology: Reflections on the Incredible Tininess of Nano." *Techné* 8, no. 3 (2005): 3–23.

Nordmann, Alfred, Joachim Schummer, and Astrid Schwarz, eds. *Nanotechnologien im Kontext: Philosophische, ethische und gesellschaftliche Perspektiven*. Berlin: Akademische Verlagsgesellschaft, 2006.

"Nothing 'Nano' about It." *BusinessWeek Online*, February 3, 2005. http://www.businessweek.com (visited April 22, 2005; web pages on file with author).

Nye, David E. *American Technological Sublime*. Cambridge: MIT Press, 1994.

Oberdörster, Eva. "Manufactured Nanomaterials (Fullerenes, C60) Induce Oxidative Stress in the Brain of Juvenile Largemouth Bass." *Environmental Health Perspectives* 112 (2004): 1058–62.

Okuda, Mike, Denise Okuda, and Debbie Mirek. *The Star Trek Encyclopedia: A Reference Guide to the Future*. New York: Pocket Books, 1994.

Ostman, Charles. "Bioconvergence: Progenitor of the Nanotechnology Age." *NanoIndustries Newsletter*, July 2000. Reprinted online at *KurzweilAI.net*. http://www.kurzweilai.net (web pages on file with author).

———. "Evolution into the Next Millennium." *Technofutures.com*, 2003. http://www.technofutures.com (visited May 20, 2004; web pages on file with author).

———. "The Nanobiology Imperative." *Technofutures.com*, 2002. http://www.technofutures.com (visited May 20, 2004; web pages on file with author).

———. "Nanobiology—Where Nanotechnology and Biology Come Together." *Biota.org*, 2003. http://www.biota.org (visited June 3, 2004; web pages on file with author).

Otis, Laura. *Membranes: Metaphors of Invasion in Nineteenth-Century Literature, Science, and Politics*. Baltimore: Johns Hopkins University Press, 1999.

Parisi, Luciana, and Steve Goodman. "The Affect of Nanoterror." *Culture Machine* 7 (2005). http://culturemachine.tees.ac.uk (web pages on file with author).

Park, Katharine. *Secrets of Women: Gender, Generation, and the Origins of Human Dissection*. Cambridge, Mass.: Zone Books, 2006.

Parker, Patricia. *Literary Fat Ladies: Rhetoric, Gender, Property*. London: Methuen, 1987.

Paschen, H., C. Coenen, T. Fleischer, R. Grünwald, D. Oertel, and C. Revermann. *Nanotechnologie: Forschung, Entwicklung, Anwendung*. Berlin: Springer-Verlag, 2004.

Pendle, George. *Strange Angel: The Otherworldly Life of Rocket Scientist John Whiteside Parsons*. Orlando: Harcourt, 2005.

Penley, Constance. *NASA/Trek: Popular Science and Sex in America*. London: Verso, 1997.

———. "Time Travel, Primal Scene and the Critical Dystopia." 1986. In *Alien Zone: Cultural Theory and Contemporary Science Fiction Cinema*, ed. Annette Kuhn, 116–27. London: Verso, 1990.

Peterson, Christine L. "Nanotechnology: Evolution of the Concept." In *Prospects in Nanotechnology: Toward Molecular Manufacturing*, ed. Markus Krummenacker and James Lewis, 173–86. New York: Wiley, 1995.

Pethes, Nicolas. "Terminal Men: Biotechnological Experimentation and the Re-shaping of 'the Human' in Medical Thrillers." *New Literary History* 36 (2005): 161–85.

Pethokoukis, James M. "The Government Says 'No' to Federally Funded Nanobots." *U.S. News and World Report* (usnews.com), December 2, 2003. http://www.usnews.com (visited April 23, 2005; web pages on file with author).

———. "Why the Feds Fear Nanobots." *U.S. News and World Report* (usnews.com), March 24, 2004. http://www.usnews.com (visited April 23, 2005; web pages on file with author).

Phoenix, Chris. "Don't Let Crichton's *Prey* Scare You—the Science Isn't Real." *Nanotechnology Now*, 2003. http://www.nanotech-now.com (visited April 23, 2005; web pages on file with author).

Phoenix, Chris, and K. Eric Drexler. "Safe Exponential Manufacturing." *Nanotechnology* 15 (2004): 869–72.

Pickering, Andrew. *The Mangle of Practice: Time, Agency, and Science*. Chicago: University of Chicago Press, 1995.

Pitt, Joseph C. "The Epistemology of the Very Small." In *Discovering the Nanoscale*, ed. Davis Baird, Alfred Nordmann, and Joachim Schummer, 157–63. Amsterdam: IOS Press, 2004.

———. "When Is an Image Not an Image?" *Techné* 8, no. 3 (2005): 24–33.

Piziks, Steven. *The Nanotech War*. New York: Simon and Schuster, 2002.

Posner, Richard. *Catastrophe: Risk and Response*. Oxford: Oxford University Press, 2004.

"Power from Blood Could Lead to 'Human Batteries.'" *Sydney Morning Herald*, August 4, 2003. http://www.smh.com.au (web pages on file with author).

Powers of Ten. Written and directed by Charles and Ray Eames. IBM, 1977.

Pressman, Jessica. "Nano Narrative: A Parable from Electronic Literature." In *Nanoculture: Implications of the New Technoscience*, ed. N. Katherine Hayles, 191–99. Bristol: Intellect Books, 2004.

Preston, Christopher J. "The Promise and Threat of Nanotechnology: Can Environmental Ethics Guide Us?" *Hyle* 11 (2005): 19–44.

Pulver, David. *GURPS Ultra-Tech 2: Hard-Core, Hard-Wired Hardware*. Ed. Gene Seabolt. Austin: Steve Jackson Games, 1999.

Radford, Tim. "Brave New World or Miniature Menace? Why Charles Fears Grey Goo Nightmare." *Guardian*, April 29, 2003, "Guardian Home Pages," 3.

Rasmussen, Nicolas. *Picture Control: The Electron Microscope and the Transformation of Biology in America, 1940–1960*. Stanford: Stanford University Press, 1997.

Rasmussen, Steen, Liaohai Chen, David Deamer, David C. Krakauer, Norman H. Packard, Peter F. Stadler, and Mark A. Bedane. "Transitions from Nonliving to Living Matter." *Science* 303, no. 5660 (2004): 963–65.

Ratner, Daniel, and Mark A. Ratner. *Nanotechnology and Homeland Security: New Weapons for New Wars*. Upper Saddle River, N.J.: Prentice Hall/PTR, 2004.

Ratner, Mark, and Daniel Ratner. *Nanotechnology: A Gentle Introduction to the Next Big Idea*. Upper Saddle River, N.J.: Prentice Hall/PTR, 2003.

Rawstern, Rocky. "Omission in the 21st Century Nanotechnology Research and Development Act." *Nanotechnology Now*, March 11, 2004. http://www.nanotech-now.com (visited April 23, 2005; web pages on file with author).

Rees, Martin. *Our Final Hour: A Scientist's Warning: How Terror, Error, and Environmental Disaster Threaten Humankind's Future in This Century—on Earth and Beyond*. New York: Basic Books, 2003.

Regan, L., and W. F. DeGrado. "Characterization of a Helical Protein Designed from First Principles." *Science* 241 (1988): 976–78.

"Regeneration." *Star Trek: Enterprise*. Episode 48. Directed by David Livingston. Written by Michael Sussman and Phyllis Strong. Paramount Television, 2003.

Regis, Ed. *Great Mambo Chicken and the Transhuman Condition: Science Slightly over the Edge*. New York: Addison-Wesley, 1990.

———. *Nano: The Emerging Science of Nanotechnology*. Boston: Little, Brown, 1995.

Reiss, Spencer. "Size Matters." *Wired* 12, no. 2 (2004): 122–23.

Relke, Diana M. A. *Drones, Clones, and Alpha Babes: Retrofitting Star Trek's Humanism, Post-9/11*. Calgary: University of Calgary Press, 2006.

Rheinberger, Hans-Jörg. *Toward a History of Epistemic Things: Synthesizing Proteins in the Test Tube*. Stanford: Stanford University Press, 1997.

Riegl, Alois. *Late Roman Art Industry*. Trans. Rolf Winkes. Rome: Giorgio Bretschneider, 1985.

Rietman, Edward A. *Molecular Engineering of Nanosystems*. New York: Springer, 2001.

Rip, Arie. "Folk Theories of Nanotechnologists." *Science as Culture* 15 (2006): 349–65.

Riviere, Joan. "Womanliness as a Masquerade." 1929. In *Formations of Fantasy*, ed. Victor Burgin, James Donald, and Cora Kaplan, 35–44. London: Methuen, 1986.

Roberts, Jody A. "Deciding the Future of Nanotechnologies: Legal Perspectives on Issues of Democracy and Technology." In *Discovering the Nanoscale*, ed. Davis Baird, Alfred Nordmann, and Joachim Schummer, 247–55. Amsterdam: IOS Press, 2004.

Robinson, Chris. "Images in Nanoscience/Technology." In *Discovering the Nanoscale*, ed. Davis Baird, Alfred Nordmann, and Joachim Schummer, 165–69. Amsterdam: ios Press, 2004.

Robinson, Kim Stanley. *Red Mars*. New York: Bantam Spectra, 1993.

Robison, Wade L. "Nano-Ethics." In *Discovering the Nanoscale*, ed. Davis Baird, Alfred Nordmann, and Joachim Schummer, 285–300. Amsterdam: ios Press, 2004.

Roco, Mihail. "International Strategy for Nanotechnology Research and Development." *Journal of Nanoparticle Research* 3 (2001): 353–60.

———. "Interview with Dr. Mihail Roco." Interview by Pamela Bailey. *NanoApex*, September 20, 2002. Formerly online at http://news.nanoapex.com (visited April 15, 2005; web pages on file with author).

———. "Nanotechnology: Convergence with Modern Biology and Medicine." *Current Opinion in Biotechnology* 14 (2003): 337–46.

———. "Nanotechnology's Future." *Scientific American* 295, no. 2 (2006): 39.

Roco, Mihail C., and William Sims Bainbridge, eds. *Converging Technologies for Improving Human Performance: Nanotechnology, Biotechnology, Information Technology and Cognitive Science*. Arlington, Va.: NSF/DOC, 2002.

———, eds. *Managing Nano-Bio-Info-Cogno Innovations: Converging Technologies in Society*. Berlin: Springer, 2006.

———, eds. *Nanotechnology: Societal Implications—Maximizing Benefit for Humanity*. Berlin: Springer, 2006.

———, eds. *Societal Implications of Nanoscience and Nanotechnology*. Boston: Kluwer Academic Publishers, 2001.

Roco, Mihail, and Renzo Tomellini, eds. *Nanotechnology: Revolutionary Opportunities and Societal Implications*. Luxembourg: Office for Official Publications of the European Communities, 2002.

Rogers, John. *The Matter of Revolution: Science, Poetry, and Politics in the Age of Milton*. Ithaca, N.Y.: Cornell University Press, 1996.

Rotman, David. "Will the Real Nanotech Please Stand Up?" *Technology Review* 102, no. 2 (1999): 47–53.

Royal Society and the Royal Academy of Engineering. *Nanoscience and Nanotechnologies: Opportunities and Uncertainties*. London: Royal Society and Royal Academy of Engineering, 2004.

Rozum, John, and J. J. Birch. "Silent Cathedrals, Part 1: The Rabbit Hole." *Xombi*, no. 1. New York: DC Comics/Milestone Comics, 1994.

Rucker, Rudy. *Postsingular*. New York: Tor, 2007.

Russell, Eric Frank. "Hobbyist." 1947. In *Major Ingredients: The Collected Short Stories of Eric Frank Russell*, ed. Rick Katze, 182–205. Framingham: NESFA Press, 2000.

Saage, Richard. "Konvergenztechnologische Zukunftsvisionen und der klassische Utopiediskurs." In *Nanotechnologien im Kontext: Philosophische, ethische und gesellschaftliche Perspektiven*, ed. Alfred Nordmann, Joachim Schummer, and Astrid Schwarz, 179–94. Berlin: Akademische Verlagsgesellschaft, 2006.

Sammon, Paul M. "Outlaws." In *Splatterpunks: Extreme Horror*, ed. Paul M. Sammon, 272–346. London: Xanadu, 1990.

Sanchez, Edward Munn. "The Expert's Role in Nanoscience and Technology." In *Discovering the Nanoscale*, ed. Davis Baird, Alfred Nordmann, and Joachim Schummer, 257–66. Amsterdam: IOS Press, 2004.

Sandhage, K. H., M. B. Dickerson, P. M. Huseman, M. A. Caranna, J. D. Clifton, T. A. Bull, T. J. Heibel, W. R. Overton, and M. E. A. Schoenwaelder. "Novel, Bioclastic Route to Self-Assembled, 3D, Chemically Tailored Meso/Nanostructures: Shape-Preserving Reactive Conversion of Biosilica (Diatom) Microshells." *Advanced Materials* 14 (2002): 429–33.

Sarewitz, Daniel, and Edward Woodhouse. "Small Is Powerful." In *Living with the Genie: Essays on Technology and the Quest for Human Mastery*, ed. Alan Lightman, Daniel Sarewitz and Christina Desser, 63–83. Washington: Island Press, 2003.

Sargent, Ted. *The Dance of Molecules: How Nanotechnology Is Changing Our Lives*. New York: Thunder's Mouth Press, 2006.

Saulnier, Beth. "Size Matters (Smaller Is Better)." *Cornell Magazine* 103, no. 4 (2001): 42–48.

Schiebinger, Londa L. *The Mind Has No Sex? Women in the Origins of Modern Science*. Cambridge: Harvard University Press, 1989.

Schiemann, Gregor. "Dissolution of the Nature-Technology Dichotomy? Perspectives from an Everyday Understanding of Nature on Nanotechnology." In *Discovering the Nanoscale*, ed. Davis Baird, Alfred Nordmann, and Joachim Schummer, 209–13. Amsterdam: IOS Press, 2004.

Schmidt, Jan C. "Unbounded Technologies: Working through the Technological Reductionism of Nanotechnology." In *Discovering the Nanoscale*, ed. Davis Baird, Alfred Nordmann, and Joachim Schummer, 35–50. Amsterdam: IOS Press, 2004.

Schneiker, C., S. Hameroff, M. Voelker, J. He, E. Dereniak, and R. McCuskey. "Scanning Tunnelling Engineering." *Journal of Microscopy* 152 (1988): 585–96.

Schuler, Emmanuelle. "Perception of Risk and Nanotechnology." In *Discovering the Nanoscale*, ed. Davis Baird, Alfred Nordmann, and Joachim Schummer, 279–84. Amsterdam: IOS Press, 2004.

Schummer, Joachim. "Interdisciplinary Issues in Nanoscale Research." In *Discovering the Nanoscale*, ed. Davis Baird, Alfred Nordmann, and Joachim Schummer, 9–20. Amsterdam: IOS Press, 2004.

———. "Multidisciplinarity, Interdisciplinarity, and Patterns of Research Collaboration in Nanoscience and Nanotechnology." *Scientometrics* 59 (2004): 425–65.

———. "Nano-Erlösung oder Nano-Armageddon? Technikethik im christlichen Fundamentalismus." In *Nanotechnologien im Kontext: Philosophische, ethische und gesellschaftliche Perspektiven*, ed. Alfred Nordmann, Joachim Schummer, and Astrid Schwarz, 263–76. Berlin: Akademische Verlagsgesellschaft, 2006.

———. "Reading Nano: The Public Interest in Nanotechnology as Reflected in Book Purchase Patterns." *Public Understanding of Science* 14 (2005): 163–83.

———. "'Societal and Ethical Implications of Nanotechnology': Meanings, Interest Groups, and Social Dynamics." *Techné* 8, no. 2 (2004): 56–87.

Schwarz, Astrid E. "Shrinking the Ecological Footprint with NanoTechnoScience?" In *Discovering the Nanoscale*, ed. Davis Baird, Alfred Nordmann, and Joachim Schummer, 203–8. Amsterdam: 1os Press, 2004.

"Scorpion." *Star Trek: Voyager*. Part 1, Episode 68: Directed by David Livingston. Part 2, Episode 59: Directed by Winrich Kolbe. Written by Brannon Braga and Joe Menosky. Paramount Television, 1997.

Seaton, Anthony, and Kenneth Donaldson. "Nanoscience, Nanotoxicology, and the Need to Think Small." *Lancet* 356, no. 9463 (2005): 923–24.

Seeman, Nadrian C. "Nanotechnology and the Double Helix." *Scientific American* 290, no. 6 (2004): 64–75.

Seeman, Nadrian C., and Angela M. Belcher. "Emulating Biology: Building Nano-structures from the Bottom Up." *PNAS* 99, suppl. 2 (2002): 6451–55.

Seltzer, Mark. *Serial Killers: Death and Life in America's Wound Culture*. New York: Routledge, 1998.

Service, Robert F. "afms Wield Parts for Nanoconstruction." *Science* 282 (1998): 1620–21.

———. "Atom-Scale Research Gets Real." *Science* 290 (2000): 1524–31.

———. "Borrowing from Biology to Power the Petite." *Science* 283 (1999): 27–28.

Shelley, Mary. *Frankenstein*. 1818. Ed. J. Paul Hunter. New York: W. W. Norton, 1996.

Shirley, John. *Crawlers*. New York: Del Rey, 2003.

Short, Harry W. "How Small Is Small?" *National Driller*, March 21, 2001. http://www.nationaldriller.com (visited November 17, 2006; web pages on file with author).

Silverberg, Robert. Introduction to *The Science Fiction Hall of Fame, Volume One, 1929–1964*, ed. Robert Silverberg, xi–xiv. New York: Doubleday, 1970. Reprint, New York: Tor Books, 2003.

Simon, Hank. "Manipulating Molecules." *Today's Chemist at Work* 10, no. 11 (2001): 36–40.

Sincell, Mark. "NanoManipulator Lets Chemists Go Mano a Mano with Molecules." *Science* 290 (2000): 1530.

Singularity Institute for Artificial Intelligence. "Nanotechnology." *Singularity Institute for Artificial Intelligence*, 2005. http://www.singinst.org (visited September 20, 2006; web pages on file with author).

Slonczewski, Joan. *Brain Plague*. New York: Tor, 2000.

———. "Stranger than Fiction." *Nature* 426 (2003): 501.

———. "Tuberculosis Bacteria Join UN." *Nature* 405 (2000): 1001.

Smalley, Richard. Foreword to *Nanocosm: Nanotechnology and the Big Changes Coming from the Inconceivably Small*, by William Illsey Atkinson. Rev. ed. New York: ama-com/American Management Association, 2005.

———. "Nanotech Growth." *R&D Magazine* 41, no. 7 (1999): 34–37.

———. "Nanotechnology: Prepared Written Statement and Supplemental Material of R. E. Smalley, Rice University, June 22, 1999." In *Nanotechnology: The State of*

Nanoscience and Its Prospects for the Next Decade; Hearing before the Subcommittee on Basic Research of the Committee on Science, House of Representatives, One Hundred Sixth Congress, First Session, June 22, 1999, 55–56. Washington: U.S. Government Printing Office, 1999.

———. "Nanotechnology, Education, and the Fear of Nanobots." In Societal Implications of Nanoscience and Nanotechnology, ed. Mihail C. Roco and William Sims Bainbridge, 145–46. Boston: Kluwer Academic Publishers, 2001.

———. "Of Chemistry, Love and Nanobots." Scientific American 285, no. 3 (2001): 76–77.

———. "Smalley Concludes." Chemical and Engineering News 81 (2003): 41–42.

———. "Smalley Responds." Chemical and Engineering News 81 (2003): 39–40.

Smarr, Larry. "Microcosmos: Nano Space." Wired 11, no. 6 (2003): 134.

Soentgen, Jens. "Atome Sehen, Atome Hören." In Nanotechnologien im Kontext: Philosophische, ethische und gesellschaftliche Perspektiven, ed. Alfred Nordmann, Joachim Schummer, and Astrid Schwarz, 97–113. Berlin: Akademische Verlagsgesellschaft, 2006.

Sound Photosynthesis. "Richard Feynman." Sound.Photosynthesis.com, 2004. http://www.photosynthesis.com (visited May 16, 2005; web pages on file with author).

Spanier, Bonnie. Im/Partial Science: Gender Ideology in Molecular Biology. Bloomington: Indiana University Press, 1995.

Spelman, Elizabeth V. "Woman as Body: Ancient and Contemporary Views." Feminist Studies 8 (1982): 109–31.

Springer, Claudia. Electronic Eros: Bodies and Desire in the Postindustrial Age. Austin: University of Texas Press, 1996.

Squier, Susan Merrill. Liminal Lives: Imagining the Human at the Frontiers of Biomedicine. Durham: Duke University Press, 2004.

Staedter, Tracy. "Sperm Power: New Tool for Nanobots." Discovery News: Discovery Channel, December 26, 2007. http://dsc.discovery.com/news (visited January 1, 2008; web pages on file with author).

Stafford, Barbara Maria. Body Criticism: Imaging the Unseen in Enlightenment Art and Medicine. Cambridge: MIT Press, 1991.

———. "Leveling the New Old Transcendence: Cognitive Coherence in the Era of Beyondness." New Literary History 35 (2004): 321–38.

Sten, Lin. Souls, Slavery, and Survival in the Molenotech Age: An Alien's Vision. St. Paul: Paragon Press, 1999.

Stephenson, Neal. The Diamond Age. New York: Bantam Books, 1995. Paperback reprint, New York: Bantam Spectra, 1996.

Stewart, Susan. On Longing: Narratives of the Miniature, the Gigantic, the Souvenir, the Collection. Durham: Duke University Press, 1993.

Stix, Gary. "Little Big Science." Scientific American 285, no. 3 (2001): 32–37.

———. "Trends in Nanotechnology: Waiting for Breakthroughs." Scientific American 274, no. 4 (1996): 94–99.

Stroscio, Joseph A., and D. M. Eigler. "Atomic and Molecular Manipulation with the Scanning Tunneling Microscope." *Science* 254 (1991): 1319–26.

Stross, Charles. *Singularity Sky*. New York: Ace Books, 2003.

Sturgeon, Theodore. "Microcosmic God." 1941. In *The Science Fiction Hall of Fame, Volume One, 1929–1964*, ed. Robert Silverberg, 88–112. New York: Doubleday, 1970. Reprint, New York: Tor Books, 2003.

Sturken, Marita, Douglas Thomas, and Sandra Ball-Rokeach. *Technological Visions: The Hopes and Fears That Shape New Technologies*. Philadelphia: Temple University Press, 2004.

Sunder Rajan, Kaushik. *Biocapital: The Constitution of Postgenomic Life*. Durham and London: Duke University Press, 2006.

Suvin, Darko. *Metamorphoses of Science Fiction: On the Poetics and History of a Literary Genre*. New Haven: Yale University Press, 1979.

Swiss Re. *Nanotechnology: Small Matter, Many Unknowns*. Zürich: Swiss Reinsurance Company, 2004.

Taniguchi, Norio. "On the Basic Concept of 'Nano-technology.'" In *Proceedings of the International Conference on Production Engineering*, part 2, 18–23. Tokyo: Japan Society of Precision Engineering, 1974.

Taylor, Russell M., Warren Robinett, Vernon L. Chi, Frederick P. Brooks Jr., William V. Wright, R. Stanley Williams, and Erik J. Snyder. "The Nanomanipulator: A Virtual-Reality Interface for a Scanning Tunneling Microscope." In *Proceedings of the 20th Annual Conference on Computer Graphics and Interactive Techniques*, ed. Mary C. Whitton, 127–34. Vol. 27, ACM SIGGRAPH — International Conference on Computer Graphics and Interactive Techniques. New York: ACM Press, 1993.

Templesmith, Ben. *Singularity 7*. San Diego: IDW Publishing, 2005.

Teresko, John. "The Next Material World." *Industry Week* 252, no. 4 (2003): 41–46.

The Terminator. Directed by James Cameron. Screenplay by James Cameron and Gale Anne Hurd. Orion, 1984.

Terminator 2: Judgment Day. Directed by James Cameron. Screenplay by James Cameron and William Wisher Jr. TriStar, 1991.

Terminator 3: Rise of the Machines. Directed by Jonathan Mostow. Screenplay by John D. Brancato and Michael Ferris. Warner, 2003.

The Texas Chain Saw Massacre. Directed by Tobe Hooper. Screenplay by Kim Henkel and Tobe Hooper. Bryanston, 1974.

Thacker, Eugene. *Biomedia*. Minneapolis: University of Minnesota Press, 2004.

———. "The Science Fiction of Technoscience: The Politics of Simulation and a Challenge for New Media Art." *Leonardo* 34, no. 2 (2001): 155–58.

Theis, Thomas N. "Letter to the Editor." *Technology Review* 102, no. 2 (1999): 15.

———. "Nanotechnology: A Revolution in the Making." *Telekom Praxis* 79, no. 12 (2002): 11–19.

Theweleit, Klaus. *Male Fantasies*. Vol. 1, *Women, Floods, Bodies, History*. 1977. Trans. Stephen Conway in collaboration with Erica Carter and Chris Turner. Minneapolis: University of Minnesota Press, 1987.

————. *Male Fantasies*. Vol. 2, *Male Bodies: Psychoanalyzing the White Terror*. 1978. Trans. Stephen Conway in collaboration with Erica Carter and Chris Turner. Minneapolis: University of Minnesota Press, 1989.

Thurs, Daniel Patrick. "Tiny Tech, Transcendent Tech: Nanotechnology, Science Fiction, and the Limits of Modern Science Talk." *Science Communication* 29 (2007): 65–95.

Timp, Gregory. "Nanotechnology." In *Nanotechnology*, ed. Gregory Timp, 1–5. New York: Springer, 1999.

————, ed. *Nanotechnology*. New York: Springer, 1999.

Todd, Dennis. *Imagining Monsters: Miscreations of the Self in Eighteenth-Century England*. Chicago: University of Chicago Press, 1995.

Tolles, W. M. "National Security Aspects of Nanotechnology." In *Societal Implications of Nanoscience and Nanotechnology*, ed. Mihail C. Roco and William Sims Bainbridge, 218–37. Boston: Kluwer Academic Publishers, 2001.

Toumey, Chris. "Apostolic Succession: Does Nanotechnology Descend from Richard Feynman's 1959 Talk?" *Engineering and Science* 68, no. 1 (2005): 16–23.

————. "The Man Who Understood the Feynman Machine." *Nature Nanotechnology* 2 (2007): 9–10.

————. "Narratives for Nanotech: Anticipating Public Reactions to Nanotechnology." *Techné* 8 (2004): 88–116.

————. "Reading Feynman into Nanotech: Does Nanotechnology Descend from Richard Feynman's 1959 Talk?" *Configurations* (forthcoming).

Traweek, Sharon. *Beamtimes and Lifetimes: The World of High Energy Physics*. Cambridge: Harvard University Press, 1988.

Treder, Mike. "Bridges to Safety, and Bridges to Progress." New York: Center for Responsible Nanotechnology, 2004. http://crnano.org (web pages on file with author).

Turner, Frederick Jackson. *The Frontier in American History*. New York: H. Holt, 1920.

Turney, Jon. *Frankenstein's Footsteps: Science, Genetics and Popular Culture*. New Haven: Yale University Press, 1998.

Ulam, Stanislaw. "John von Neumann, 1903–1957." *Bulletin of the American Mathematical Society* 64, no. 3, pt. 2 (1958): 1–49.

Uldrich, Jack, and Deb Newberry. *The Next Big Thing Is Really Small: How Nanotechnology Will Change the Future of Your Business*. New York: Crown Business, 2003.

United States Congress. House of Representatives. Committee on Science. *H.R. 766: Nanotechnology Research and Development Act of 2003; Hearing before the Committee on Science, House of Representatives, One Hundred Eighth Congress, First Session, March 19, 2003*. Washington: U.S. Government Printing Office, 2003.

————. *Nanotechnology: The State of Nanoscience and Its Prospects for the Next Decade; Hearing before the Subcommittee on Basic Research of the Committee on Science,*

House of Representatives, One Hundred Sixth Congress, First Session, June 22, 1999. Washington: U.S. Government Printing Office, 1999.

——. *The Societal Implications of Nanotechnology: Hearing before the Committee on Science, House of Representatives, One Hundred Eighth Congress, First Session, April 9, 2003.* Washington: U.S. Government Printing Office, 2003.

United States Congress. Senate. Committee on Commerce, Science, and Transportation. *New Technologies for a Sustainable World: Hearing before the Subcommittee on Science, Technology, and Space of the Committee on Commerce, Science, and Transportation, United States Senate, One Hundred Second Congress, Second Session, June 26, 1992.* Washington: U.S. Government Printing Office, 1993.

——. *21st Century Nanotechnology Research and Development Act: Report of the Committee on Commerce, Science, and Transportation on S. 189.* Washington: U.S. Government Printing Office, 2003.

van Lente, Harro, and Arie Rip. "The Rise of Membrane Technology: From Rhetorics to Social Reality." *Social Studies of Science* 28 (1998): 221–54.

Varela, Francisco J. "Organism: A Meshwork of Selfless Selves." In *Organism and the Origins of Self*, ed. Alfred I. Tauber, 79–107. Dordrecht: Kluwer Academic Publishers, 1991.

Vermaas, Pieter E. "Nanoscale Technology: A Two-Sided Challenge for Interpretations of Quantum Mechanics." In *Discovering the Nanoscale*, ed. Davis Baird, Alfred Nordmann, and Joachim Schummer, 77–91. Amsterdam: 10s Press, 2004.

Videodrome. Directed by David Cronenberg. Screenplay by David Cronenberg. Universal, 1983.

Vinge, Vernor. "'Bookworm, Run!'" 1966. In *The Collected Stories of Vernor Vinge*, 15–44. New York: Tor Books, 2001.

——. *A Fire upon the Deep.* New York: Tor Books, 1992.

——. "First Word." *Omni* 5 (January 1983): 10.

——. *Marooned in Realtime.* 1986. New York: Tor Books, 2004.

——. *The Peace War.* 1984. New York: Tor Books, 2003.

——. "Technological Singularity." *Whole Earth Review* 81 (1993): 88–95.

Vint, Sherryl. *Bodies of Tomorrow: Technology, Subjectivity, Science Fiction.* Toronto: University of Toronto Press, 2007.

Virilio, Paul. *The Art of the Motor.* Trans. Julie Rose. Minneapolis: University of Minnesota Press, 1995.

——. *The Vision Machine.* Trans. Julie Rose. Bloomington: Indiana University Press, 1994.

——. *War and Cinema: The Logistics of Perception.* 1984. Trans. Patrick Camiller. London: Verso, 1989.

Voss, David. "Moses of the Nanoworld." *Technology Review* 102, no. 2 (1999): 60–62.

Waddington, Simon. "Grußwort/Preface: Nanotechnologie 2002." In "Nanotechnologie," special issue, *VentureCapital Magazin*, October 2002, 4.

Wald, Carol Ann. "Working Boundaries on the *nano* Exhibition." In *Nanoculture: Implications of the New Technoscience*, ed. N. Katherine Hayles, 83–104. Bristol: Intellect Books, 2004.

Waldby, Catherine. *The Visible Human Project: Informatic Bodies and Posthuman Medicine*. London: Routledge, 2000.

Weinstone, Ann. *Avatar Bodies: A Tantra for Posthumanism*. Minneapolis: University of Minnesota Press, 2004.

Welland, M., and J. K. Gimzewski, eds. *The Ultimate Limits of Fabrication and Measurement*. Dordrecht: Kluwer Academic Publishers, 1995.

Whitesides, George M. "Nanotechnology: Art of the Possible." Interview by David Rotman. *Technology Review* 101, no. 6 (1998): 84–87.

———. "The New Biochemphysicist." Interview by David Ewing Duncan. *Discover* 24 (December 2003): 24.

———. "The Once and Future Nanomachine." *Scientific American* 285, no. 3 (2001): 78–83.

———. "Self-Assembling Materials." *Scientific American* 273, no. 3 (1995): 146–49.

Whitesides, George M., and Mila Boncheva. "Beyond Molecules: Self-Assembly of Mesoscopic and Macroscopic Components." *PNAS* 99, no. 8 (2002): 4769–74.

Whitesides, George M., John P. Mathias, and Christopher T. Seto. "Molecular Self-Assembly and Nanochemistry: A Chemical Strategy for the Synthesis of Nanostructures." *Science* 254 (1991): 1312–19.

Wiener, Norbert. *The Human Use of Human Beings: Cybernetics and Society*. Boston: Houghton Mifflin, 1954.

Williams, Linda. *Hard Core: Power, Pleasure, and the "Frenzy of the Visible."* 1989. 2nd ed. Berkeley: University of California Press, 1999.

Willis, Martin. *Mesmerists, Monsters, and Machines: Science Fiction and the Cultures of Science in the Nineteenth Century*. Kent, Ohio: Kent State University Press, 2006.

Winner, Langdon. *Autonomous Technology: Technics-out-of-Control as a Theme in Political Thought*. Cambridge: MIT Press, 1977.

Winsberg, Eric. "Handshaking Your Way to the Top: Simulation at the Nanoscale." In *Simulation: Pragmatic Construction of Reality*, ed. Johannes Lenhard, Günter Küppers, and Terry Shinn, 139–51. Dordrecht: Springer, 2006.

Wolfe, Cary. *Animal Rites: American Culture, the Discourse of Species, and Posthumanist Theory*. Chicago: University of Chicago Press, 2003.

Wolfe, Josh. "Top Nano Products of 2005." *Forbes/Wolfe Nanotech Report* 4, no. 12 (2005): 1–2. http://www.forbes.com (web pages on file with author).

Yakobson, Boris I., and Richard Smalley. "Fullerene Nanotubes: $C_{1,000,000}$ and Beyond." *American Scientist* 85, no. 4 (1997): 324–37.

Žižek, Slavoj. *Enjoy Your Symptom! Jacques Lacan in Hollywood and Out*. 1992. Rev. ed. New York: Routledge, 2001.

———. *The Sublime Object of Ideology*. New York: Verso, 1989.

INDEX

Barker, Clive, 168, 228 n. 56
Batman, 44
Baudrillard, Jean, 25–26, 40, 79
Baxter, Stephen, 107, 214 n. 130
Bear, Greg, 51, 178–182, 183, 229 n. 70
"Beauty and Brains" (GE), 136, 151–158, 224 n. 110
becoming, 15–16, 25, 56–58, 92, 144, 151, 160, 162, 184–186, 192 n. 40, 193 n. 42, 203 n. 112, 222 n. 96; nanotechnology impact on, 50–54, 184–186; redefinition of life and, 164–167, 175–178
Benford, Gregory, 4, 42, 44, 200 n. 86
Benjamin, Walter, 81, 210 n. 74
Berger, John, 192 n. 36
Berne, Rosalyn W., 201 n. 93, 215 n. 3
Beyond the Pleasure Principle (Freud), 228 n. 59
Biagioli, Mario, 205 n. 16
Binnig, Gerd, 30–31, 69, 80–83, 85–87, 92, 95, 108–109, 122–123
biology, 6, 45, 51, 56, 71, 161–170, 173–178, 183–186, 224 n. 1, 226 n. 21, 227 n. 22, 227 n. 38, 229 n. 70
bionanotechnology. *See* nanobiology
biotechnology, 12, 45, 161–165, 215 n. 5, 216 n. 23, 224 n. 1, 226 n. 17
black holes, 2–3, 12–13, 222 n. 81
Blade Runner, 53
blindness, 2–5, 9–18, 35, 50, 86–88, 94, 99–100, 102, 123–128, 141, 154–160, 184–187. *See also* vision
Blish, James, 47, 96–106, 107, 213 n. 118, 213 n. 126
Block, Steven M., 23
Blohm, Margaret, 152
Blood Music (Bear), 51, 178–182, 229 n. 70
Bloom (McCarthy), 15–16, 134–135, 158–161, 163, 173–175, 178, 181–184
Bloom, Harold, 207 n. 39

blue goo, 142–143. *See also* gray goo; green goo
body count, 169, 170
"body without organs," 185–186, 228 n. 43
Boehlert, Sherwood, 126
Bolter, Jay David, 90
Brin, David, 44, 200 n. 86
Bristol Centre for Nanoscience and Quantum Information (U.K.), 218 n. 50
Brite, Poppy Z., 168
Broderick, Damien, 3–6
Brown, Nathan, 197 n. 53, 206 n. 33
Brownian motion, 80, 185
buckminsterfullerenes, 30–31, 42, 132–133, 218 n. 50
buckyballs. *See* buckminsterfullerenes
buckytubes. *See* nanotubes
Bukatman, Scott, 27, 204 n. 117, 222 n. 89
Burson-Marsteller, 116
Bush, George W., 20
Bush, Vannevar, 78–79
Butler, Judith, 129–130, 221 n. 74

Calder, Richard, 168
California Institute of Technology, 36, 38, 198 n. 59, 201 n. 98
Cameron, James, 136–137
Cannon, Walter B., 226 n. 22
Čapek, Karel, 44
Carroll, Lewis, 79, 209 n. 66
Center for Integrated Nanotechnologies, 60, 79
Center for Responsible Nanotechnology, 21, 44, 112, 200 n. 86
Chaga. See *Evolution's Shore*
Charles, Prince of Wales (Charles Windsor), 127–128
chemistry, 6, 19, 30, 71, 83, 85, 88, 94–95, 126, 127, 150, 166, 179, 197 n. 48, 217 n. 28, 219 n. 53, 229

n. 64. *See also* supramolecular chemistry

Clark, Andy, 211 n. 101

Clarke, Arthur C., 41, 200 n. 80; *The Fountains of Paradise*, 42–43; *Profiles of the Future*, 59, 199 n. 68

Clement, Hal, 47

Clinton, William J., 20, 37–38, 72

cloning, 6, 49, 52–53, 116

Clover, Carol, 228 n. 41

cognitive estrangement, 24, 27, 45, 79–80, 189 n. 3

computing, 1, 3–5, 20, 23, 28, 41, 46, 53, 61, 82, 84–85, 135, 136, 144, 165, 190 n. 17, 191 n. 20, 196 n. 38; quantum, 6, 9

Conley, Tom, 205 n. 25

converging sciences, 9, 12, 32, 34, 45, 50, 192 n. 32, 197 n. 53, 199 n. 75, 215 n. 5

Cornell University, 204 n. 119, 222 n. 83

Cram, Donald, 30

Cramer, John G., 22, 44, 111, 200 n. 86

Crandall, B. C., 22, 166

Crash (Ballard), 88

Crawlers (Shirley), 163, 169, 172–173, 183

Creed, Barbara, 131, 222 n. 81

Crichton, Michael: *Disclosure*, 135; *Prey*, 118, 135–136, 175–176, 217 n. 31; *Rising Sun*, 135

Crick, Francis, 164

Cronenberg, David, 168

cryonics, 29, 54–58, 62, 202 n. 103, 203 nn. 112–113, 203 n. 116

Csicsery-Ronay, Istvan, 195 n. 29

Culler, Jonathan, 42

Cummings, Ray, 47, 94–95, 213 n. 114

Curl, Robert, 30

cybernetics, 5, 11, 25, 45, 49, 138, 144, 172

cyborgs, 5, 27, 29, 45, 49–52, 54–56, 87–90, 136–146, 172, 196 n. 38, 211 n. 101, 222 n. 90, 223 nn. 96–97

Daar, Abdallah, 216 n. 23

Daston, Lorraine, 210 n. 75

Day the Earth Stood Still, The, 1

deconstruction, 14, 16–17, 25–27, 38–49, 107, 122, 130, 168–169, 184, 192 n. 35, 201 n. 99, 221 n. 79, 229 n. 70. *See also* techno-deconstruction

DeGrado, William, 30

De Landa, Manuel, 166, 227 n. 27, 227 n. 36

Deleuze, Gilles, 85, 165–166, 175, 180–181, 184–186, 192 n. 38, 228 n. 43

de Man, Paul, 192 n. 35

Democritus, 193 n. 8

Derrida, Jacques, 49, 98–99, 120, 192 n. 38, 201 n. 99, 207 n. 39, 221 n. 78, 225 n. 6

Dery, Mark, 14, 202 n. 104, 222 n. 89

desiring-machines, 181–182, 185

de Souza e Silva, Adriana, 202 n. 100

destining, 11, 22, 61–67, 70, 130, 146, 154, 162–167, 175–177, 186, 221 n. 78, 225 n. 6, 229 n. 70

diamond, 7, 20, 29, 52, 61, 213 n. 114

Diamond Age, 42, 57

Diamond Age, The (Stephenson), 19, 42, 51, 58

"Diamond Lens, The" (O'Brien), 213 n. 114

diatom nanotechnology, 43

Diaz-Rahi, Yamila, 152

Dick, Philip K., 47

Disclosure (Crichton), 135

disintegration, 17, 125, 131, 147, 149, 162–184, 225 n. 6, 228 n. 43, 229 n. 64

Disney, Walt, 56–62, 203 n. 116

Disneyism, 56–57, 60, 85, 204 n. 117

Disneyland, 56–58

NanoBusiness Alliance (U.S.), 116, 124
nanoconvergence. *See* converging
  sciences
Nanodreams, 39
"Nano energy underclothes," 11
nanofactories, 20, 46, 47, 112
nanofinger, 85–87, 100. *See also* nanotechnology: as shaped by tropes
nanofuture, 12, 15, 22, 27–28, 32,
  37–38, 40–45, 57–58, 70–71, 79, 110,
  112–118, 124–126, 129, 152, 163–169,
  173–177, 186–187, 193 n. 8, 195 n. 36,
  204 n. 121, 217 n. 23
Nano Letters, 33
nanomachines, 1, 6–7, 19–20, 23, 24,
  29, 43, 45, 52–54, 56, 95, 119, 122,
  128, 132, 142–143, 164–165, 183, 185,
  194 n. 12, 197 n. 48, 220 n. 58, 226
  n. 21; disintegration and refabrica-
  tion and, 169–173, 176–178; self-
  replication of, 114–115, 117–118, 126,
  173
nanoManipulator (UNC, Chapel Hill),
  83–85, 91, 100, 212 n. 107
nanomedicine, 20, 21, 22, 29, 34, 43,
  46, 47, 48, 55–56, 57, 95, 112, 124,
  161, 162, 165, 196 n. 38, 226 n. 17
nanonarratives, 39, 45, 50–53, 56, 113,
  118, 163–164, 168–186, 215 n. 1, 228
  n. 59
"Nanoprobe" (Jay), 7
nanorhetoric, 9, 24, 26–38, 62–63, 66,
  86, 116, 123, 147–151, 163, 177, 194
nanoscale, 6, 8–9, 12, 20, 30, 34,
  43–45, 50, 56, 70–73, 107–109, 122,
  161–162, 186, 191 n. 20, 206 n. 26,
  207 n. 42, 210 n. 78, 224 nn. 1–2;
  frontier of, 73–83, 92–96; visualization and imaging, 62–69, 91–92
nanoscience. *See* nanotechnology
nanosplatter, 175–176, 178, 180–184.
  See also disintegration
Nanosystems (Drexler), 33, 40, 194 n. 12

Nanotech Chronicles, The (Flynn), 54
nanotechnological transjectors, 138–
  143
nanotechnology, 6–9, 19–20, 197
  n. 56; comparison to biological
  systems, 19, 29, 41, 51, 170, 176–177;
  as complicit with science fiction,
  23–28, 39–49, 84–85, 118, 194 n. 12,
  196 n. 38, 201 n. 98, 219 n. 53; as
  control of matter, 19–20, 28, 62–72,
  91–92, 104–110, 112–114, 166–168,
  205 n. 18; critiques of, 23–24, 111–
  115; cyborgs and, 27, 52, 132; domestic security and, 146–158; dry versus
  wet, 168, 224 n. 1; ethics and, 15,
  73, 106–112, 129, 164, 167, 184, 216
  n. 23; futuristic visions of, 6, 16, 20–
  23, 39–49; governmental funding
  initiatives for, 8, 20–21, 116–117; as
  hard science, 126–129, 220 nn. 70–
  71; history of, 8, 30–34; and the
  human, 128–130; as hyperreal, 26,
  44; maternal fantasies and, 116, 119–
  121, 129–136; militarism and, 146–
  151; as multidisciplinary research
  field, 6, 32–33, 197 n. 47; opposed
  to science fiction, 23, 26, 28–38,
  115, 118, 125–126, 219 n. 53; pedagogy and, 33; and the posthuman,
  13–18, 27, 49–58, 185–187; relation
  to nanovision, 9–14, 61–74; relation
  to Singularity, 6–8, 10, 190 n. 17;
  as represented in science fiction, 1,
  15–19, 44–45, 47, 62, 96–107, 118,
  132–146, 168–187; risks of, 111–116,
  124, 128, 216 n. 23, 221 n. 71; as
  shaped by tropes of touching, fingering, handling, 10, 11, 16, 38, 48, 56,
  85–93, 95–96, 99–100, 103, 108,
  156–159, 162 (*see also* haptic vision);
  as site of converging sciences, 9, 32,
  34; as technological revolution, 9–11,
  28, 50; in textbooks, 33, 42; as trans-

phallus, 100, 120, 123, 130, 137, 141, 144, 146, 154, 221 n. 78, 222 n. 88. *See also* technoscientific phallus

Phoenix, Chris, 117

physics, 2–3, 4, 6, 8, 22, 23, 32, 36–37, 44, 69, 71, 93, 127, 179, 206 n. 32, 221 n. 74

Pickering, Andrew, 167

"Pirates of the Caribbean" (Disneyland), 56

Pitt, Joseph, 210 n. 78

Piziks, Steven, 43

postbiological life, 5, 162–167, 169–187, 214 n. 129, 225 n. 13, 228 n. 59, 229 n. 64

post effect (Derrida), 186, 207 n. 39, 225 n. 6

posthuman, 2–5, 14–16, 27–28, 103; embodiment and, 50–58, 88–92, 183–187, 202 n. 104; as engineered by nanovision, 16–18, 49–58, 80–92, 109–110, 128–129, 160, 176–177; as love, 109, 184–187; psychic apparatus of, 90–93; in relation to humanism, 193 n. 42, 193 n. 45, 229 n. 70; sensorium of, 87–89; sexuality and, 53, 108–110, 139–141, 222 nn. 89–97

Postsingular (Rucker), 1, 190 n. 17

Powers of Ten, 213 n. 117

Pressman, Jessica, 191 n. 20

Prey (Crichton), 118, 135–136, 175–176, 217 n. 31

Profiles of the Future (Clarke), 59, 199 n. 68

psychoanalysis, 89–90, 129, 131, 221 n. 76

Pytka, Joe, 151, 152

quantum computing, 6, 9

quantum corral, 65–67

quantum dots, 32

quantum entanglement, 6, 90, 95

quantum theory, 24, 33, 62, 65–68, 71–72, 80, 84, 86, 90, 92, 108, 178–180, 212 n. 110

quantum tunneling. *See* tunneling

Quate, Calvin, 30

Queen City Jazz (Goonan), 51, 163

rapture. *See* eschatology

Rasmussen, Nicolas, 81, 209 n. 72, 211 n. 100

Rasmussen, Steen, 164

Ratner, Daniel, 9, 136, 146–151, 223 n. 102

Ratner, Mark A., 9, 136, 146–151, 223 n. 102

Rayleigh limit, 71–72, 208 n. 48

Reed, Mark, 36, 40–41, 46

Reitman, Edward, 42

Rheinberger, Hans-Jörg, 79

Rice University, 199 n. 74

Riegl, Alöis, 85

Rip, Arie, 194 n. 12, 216 n. 20

Rising Sun (Crichton), 135

Riviere, Joan, 141

Robinett, Warren, 84–85

Roco, Mihail, 20, 30, 115, 193 n. 2, 196 n. 38, 219 n. 54, 219 n. 56, 224 n. 1

Rohrer, Heinrich, 30–31, 69, 80, 82, 85–87, 92, 95, 108–109, 122–123, role-playing games, 155, 169

Romero, George, 168

Royal Society and the Royal Academy of Engineering, 112

Rucker, Rudy, 1, 190 n. 17

Ruska, Ernst, 30

Russell, Erik Frank, 47

Sanchez, Edward Munn, 220 n. 69

Sargent, Ted, 28, 123, 166–167

scanning near-field optical microscopy (SNOM), 208 n. 48

scanning probe microscopy (SPM), 19, 29, 62, 63, 69, 70, 74, 76, 80, 95, 100, 108–109, 123, 191 n. 20, 208 n. 48, 209 n. 70, 211 n. 100; compared to atomic force microscope, 30; experience of, 82–91. *See also* scanning tunneling microscopy

scanning tunneling microscopy (STM), 30–31, 62–93, 94, 95, 100, 122–123, 205 n. 17, 207 n. 42, 210 n. 78, 212 n. 110

scapegoat, 125–126

scene of disintegration. *See* disintegration

Schmidt, Jan, 192 n. 32

Schneiker, Conrad, 69, 207 n. 36

Schuler, Emmanuelle, 116

Schummer, Joachim, 190 n. 18, 192 n. 39, 194 n. 12, 201 n. 93, 216 n. 23, 224 n. 102

Schwarzenegger, Arnold, 136, 138, 140, 146, 222 n. 88

Schweizer, Erhard, 30–31, 64–65, 205 n. 18

Science Citation Index, 199 n. 72

science fiction: formal characteristics of, 24, 79, 209 n. 68; hyperreality and, 25–27; as opposed to real science, 8, 13, 23–26, 28–29, 34–38, 115, 125–126; representations of nanotechnology in, 1, 15–19, 44–45, 47, 62, 96–107, 118, 132–146, 168–187; sense of wonder and, 46, 79, as shaping real science, 24–28, 37, 39–49, 84–85, 118, 168–187; writers of, 1–2, 11–12, 35, 39, 41, 44–46, 47, 95–96, 132–136, 168–187

Science Fiction Hall of Fame, The, 213 n. 118, 214 n. 129

Science Fiction Writers of America (SFWA), 96, 214 n. 129

Sciencenter, 204 n. 119

Seeman, Nadrian, 164

self-assembly, 19, 80, 113, 114, 125, 164–166, 218 n. 53,

self-organization, 113, 114, 121, 124, 131, 164–167, 175, 222 n. 80, 228 n. 57

self-replication, 23, 40, 41, 47, 52, 61, 113–117, 122, 124–126, 129, 132, 142, 146, 164–169, 173, 177, 190, 197 n. 49, 218 n. 53, 223 n. 102, 225 n. 2, 226 nn. 18–19, 226 n. 22

Seltzer, Mark, 172

sense of wonder, 46, 79–80, 209 n. 68

Service, Robert F., 21–22

Shelley, Mary, 114, 117, 119

Shirley, John, 163, 168, 169, 172–173, 183

Shrinking Man, The (Matheson), 96

Silicon Valley, 34

simulacrum, 25–26, 34, 39, 53, 153–155

simulation, 5, 14, 23, 25–26, 28, 29, 39, 43–44, 49, 56–57, 59, 82, 155, 163, 191 n. 20, 219 n. 53

Singer, Peter, 216 n. 23

Singularity, 1–6, 9–18, 50, 60, 74, 78, 133, 162, 181, 187

Singularity Institute for Artificial Intelligence, 190 n. 17

Singularity 7 (Templesmith), 16–17, 169

Singularity Sky (Stross), 190 n. 17

Slonczewski, Joan, 44–45

Smalley, Richard, 9, 22, 28, 30, 32, 34, 37, 40, 42, 74, 115–118, 120, 197 n. 48, 199 n. 74, 200 n. 80, 219 n. 54, 224 n. 1

Small Times, 21

Smith, Hamilton, 225 n. 2

soldier male, 131–132, 146–151

space elevator, 42–43, 200 n. 80

spaceships, 20, 28–29, 35, 45, 102–104, 134–135

Smarr, Larry, 62

Sound Photosynthesis, 36–37, 38

splatter, 138, 160, 168–169, 171, 173,

COLIN MILBURN

is an assistant professor of English and
a member of the Science and Technology
Studies program at the University of
California, Davis.

LIBRARY OF CONGRESS

CATALOGING-IN-PUBLICATION

DATA

Milburn, Colin

Nanovision : engineering the future /
Colin Milburn.

p. cm.

Includes bibliographical references and index.

ISBN 978-0-8223-4243-4 (cloth : alk. paper)

ISBN 978-0-8223-4265-6 (pbk. : alk. paper)

1. Nanoelectronics. 2. Nanotechnology.

3. Technological forecasting. I. Title.

TA418.9.N35M56 2008

620'.5—dc22 2008028433